MW00465746

SYSTEMS ENGINEERING GUIDEBOOK

A Process for Developing Systems and Products

systems engineering series

series editor
A. Terry Bahill, University of Arizona

Systems Engineering Guidebook: A Process for Developing Systems and Products
James N. Martin, Texas Instruments

The Art of Systems Architecting
Eberhardt Rechtin, University of Southern California
Mark W. Maier, University of Alabama at Huntsville

Fuzzy Rule-Based Modeling with Applications to Geophysical, Biological and Engineering Systems
András Bárdossy, University of Stuttgart
Lucien Duckstein, University of Arizona

Systems Engineering Planning and Enterprise Identity
Jeffrey O. Grady, JOG System Engineering

Systems Integration
Jeffrey O. Grady, JOG System Engineering

Model-Based Systems Engineering
A. Wayne Wymore, Systems Analysis and Design Systems

Linear Systems Theory
Ferenc Szidarovszky, University of Arizona
A. Terry Bahill, University of Arizona

The Road Map to Repeatable Success: Using QFD to Implement Change
Barbara A. Bicknell, Bicknell Consulting, Inc.
Kris D. Bicknell, Bicknell Consulting, Inc.

Engineering Modeling and Design
William L. Chapman, Hughes Aircraft Company
A. Terry Bahill, University of Arizona
A. Wayne Wymore, Systems Analysis and Design Systems

The Theory and Applications of Iteration Methods
Ioannis K. Argyros, Cameron University
Ferenc Szidarovszky, University of Arizona

System Validation and Verification
Jeffrey O. Grady, JOG System Engineering

SYSTEMS ENGINEERING GUIDEBOOK

A Process for Developing Systems and Products

James N. Martin

CRC Press

Boca Raton Boston London New York Washington, D.C.

Library of Congress Cataloging-in-Publication Data

Martin, James N.
  Systems engineering guidebook: a process for developing systems and products / by
  James N. Martin
      p. cm. — (Systems engineering series)
  Includes bibliographical references and index.
  ISBN 0-8493-7837-0
  1. Systems engineering.   I. Title. II. Series.
TA168.M267 1996
620'.001'1—dc20   96-36435
                                                                                          CIP

© 2000 by CRC Press LLC

No claim to original U.S. Government works
International Standard Book Number 0-8493-7837-0
Library of Congress Card Number 96-36435
Printed in the United States of America   4 5 6 7 8 9 0
Printed on acid-free paper

About the Author

James N. Martin holds a B.S. in Mechanical Engineering from Texas A&M University and an M.S. in Engineering from Stanford University. He is a founding member of the International Council on Systems Engineering (INCOSE) and has served on its board of directors. He is currently leading a working group to develop the EIA standard on systems engineering.

He started his engineering career with AT&T Bell Labs working on cellular wireless telecommunications. Later, he was a systems engineer on underwater fiber optic transmission systems. Mr. Martin recently joined Texas Instruments as a systems engineer and program manager for telecommunications systems.

Contents

List of Figures

List of Tables

Preface

"Writing a book is an adventure; to begin with it is a toy
and an amusement, then it becomes a master, and then
it becomes a tyrant; and...just as you are about to be
reconciled to your servitude—you kill the monster and
fling him...to the public."

—Winston Churchill

This book is different from most books on systems engineering.
Whereas most books on this subject expound upon *methods* for
analyzing performance, defining requirements and architecture, or
performing specialty engineering tasks such as reliability, life cycle
cost, and supportability, this book describes a *process framework* for
implementing the methods of *engineering a system*.

First, it is not a "cookbook," some chronological step-by-step
instruction manual on how to create the "perfect" system. There is no
perfect system, and probably never will be. This book contains rather
a blueprint that offers a solid model backed by commonsense best
practices from leading companies all over the world on how to
effectively and efficiently develop systems that satisfy the end user and
all other stakeholders.

Second, this book is not about what systems engineers do. It is
about what engineers need to do if they are dealing with complex
problems, complex technologies, or unprecedented solutions. It is
about organizing the information that is needed to develop the right
solutions and communicating that information to all affected parties in
and around a development project.

Third, it is about understanding the problem before designing the
solution. Most engineers come out of the university all geared up to
solve the problems of the world. It is a rude shock for them to
discover that the problems they are to solve have not even been

defined yet. There are no "professors" in the real world to neatly spell out the problem for them. Often, we end up with engineered solutions looking for a problem. The systems engineering process is structured in such a way as to facilitate problem definition.

Fourth, complex systems cannot be developed without involving all relevant disciplines—technical and non-technical—up-front in the development process. Proper execution of this process demands this involvement by the various "stakeholders" of a system. You will see throughout this book an emphasis on integrated product development using integrated product teams.

Fifth, systems engineering is more a way of thinking than a discipline. Granted there are special skills and abilities needed to be able to deal with the more abstract notions at higher levels in a system architecture, but this is no different from the need for special skills and abilities at the *lowest* levels of this architecture (e.g., material science, connector design, circuit board layout, laser optics, communications protocol, assembly language). Systems engineering is not merely another branch of engineering like mechanical engineering, electrical engineering, civil engineering, etc. Instead, it is a special way of looking at the world, using the "systems approach." Some people might be born with this "world view," but many people can acquire this frame of mind by practicing the "art" of systems engineering. It is difficult to describe any art per se. Hence, the process defined in this book merely represents an instantiation of the systems principles and concepts. Just like the painted canvas in the museum, the real art is somewhere below the surface of what you actually see. Most artists require much study and practice before they become proficient and well accomplished. The same goes for engineers.

Finally, this book does not represent the final say on what the process should be for the engineering of systems. This field is changing and growing. Systems are getting more complex. We know that technology does not provide the answers to all the problems of the world. There is much to learn about how to best design solutions to complex problems while at the same time giving due consideration to the needs of the world at large—environmental balance, social harmony, economic stability.

It is my hope that you will gain an appreciation for the systems approach through reading this book. I also hope that you will gain respect for the *real* issue in developing systems—people. Any process can appear to be sterile, devoid of flexibility, and an inhibitor to creativity. Quite the contrary. A process, if properly implemented, will greatly enhance the application of creativity to the *right problem*. It does little good to have the right solution for the wrong problem.

This book describes the Systems Engineering process used within AT&T Bell Labs. The material for this book was originally developed as an aid in developing products for the Advanced Technology Systems sector of AT&T. Even though the original document was intended for use in developing products for the government, it has been found useful for commercial products as well. The SE process contained herein was designated as a Best Current Practice for use throughout AT&T.

Chapter 1 provides some background for systems engineering. It describes some key elements and key issues of systems engineering. Chapter 2 defines some of the key systems terminology and concepts necessary in understanding how to apply the process. Chapter 3 describes the PMTE paradigm which will help you understand the relationship between process, methods, tools, and environment. Chapter 4 gives an overview of the process along with a description of some of the key elements such as the Systems Engineering Management Plan (SEMP), the Systems Engineering Master Schedule (SEMS), and Systems Engineering Detailed Schedule (SEDS).

Chapters 5 through 8 describe details of the Systems Engineering Process. Chapters 9 and 10 give guidance on process tailoring and process support. Chapter 11 provides guidance on application of the process in a project environment. A Glossary is provided which defines many of the unique terms related to systems engineering and a Bibliography lists many of the books and other documents relevant to its practice.

I welcome any and all questions or comments you have, or suggestions for improvement of this material. Please send your thoughts to jmartin@airmail.net.

James N. Martin
Dallas, Texas
September 1996

Acknowledgments

Most of the material in this book was originally developed as part of a Total Quality Management (TQM) effort in the Advanced Technology Systems sector of AT&T. The AT&T process document went through two editions prior to being used as source material for this book.

Jim Armstrong and George Hudak were the primary co-authors I worked with in developing the original material for this book. They provided tremendous insights into the true nature of systems engineering and they provided a wealth of experience.

I would like to especially thank the many people who provided me with the education and training in the fundamentals of systems engineering: Jim Armstrong, Brian Mar, Jerry Lake, Richard Stevens, Ben Blanchard and Troy Caver, among others. I have also had the pleasure of working with many bright engineers in the International Council on Systems Engineering (INCOSE) and on the working group responsible for developing the EIA standard on systems engineering.

Celina Rios Mullen provided the spark in my heart for the written word. Ron Frerking provided many hours of thoughtful conversation on the philosophical basis for the systems approach. Gerald Weinberg provided through his books my education in systems science and systems theory. Arthur Hall through his 1962 book, *A Methodology for Systems Engineering*, gave me hope that this thing called systems engineering was actually worth learning about. Eberhardt Rechtin provided many useful concepts with respect to systems architecting.

Jim Armstrong provided me with my first real "training" in the practices and principles of systems engineering. Don McKinney ingrained in me a deep respect for the needs of the customer and convinced me that systems engineering is really nothing but common sense (albeit not as common as it should be). Joe Punturieri made it all look so easy. He was a natural when it came to doing good systems engineering.

Ron Wade and Terry Bahill both provided many useful comments in their review of early drafts of the manuscript. Special thanks go to the members of my EIA technical committee that is developing the updated standard on systems engineering: Jerry Lake, Richard Harwell, and John Velman. We have had countless hours of discussion and hard work in getting a clear and consistent understanding of the principles and concepts of SE. They also provided much of the material for Chapter 2.

Of course, I cannot thank my wife enough for the loving support she provided during the countless hours I was missing from the dinner table. She was also my editor-in-chief, giving me many useful comments as she reviewed portions of the manuscript.

Below are the acknowledgments from the two internal editions of the AT&T process document.

1st Edition of AT&T Document

The Systems Engineering Process Management Team (PMT) chartered a Top-Level Process Definition Quality Improvement Team (QIT) to develop this document and its supporting material. James Martin led the QIT and his team members were George Aitken, Jim Armstrong, John Beck, Howard Bodner, Jerry Halpern, Lance Hancock, Bob Heinzerling, George Hudak, Bruce Porter, Glenn Marum, Wendell Miyaji, Joe Punturieri, Ron Rapp, and John Rossi. This document was written primarily by James Martin, Jim Armstrong, Lance Hancock, and George Hudak.

The Systems Engineering PMT consists of Dick Tiel (process owner), Bruce Barrett, Karen Davis, Clark DeHaven, Bryson Epting, Mike Geller, George Hudak, Don McKinney, Gene Scarborough, Dick Stratton, and John Wronka.

The Systems Engineering PMT wishes to thank the many people who participated in document reviews and deployment trials. The PMT also wishes to thank Troy Caver and his Systems Management and Development Corporation for their valuable insights and for development of the training courses associated with the process.

2nd Edition of AT&T Document

The Editor-in-Chief for the second edition was James Martin. The process metrics update team consisted of Bill Miller, Dick Stratton, Allan Tubbs, and Jim Stekas. The SI&V update team consisted of Gary Hallock, Norman File, Don Grell, F. T. Kaplysz, Jeff Kurt, Donnie Price, and Mike Wilkey. Other people who contributed to this version of the document are Joe Punturieri, Tom Uhrig, Mary Ann Welsh, Gene Scarborough, Yan Shi, Wendell Miyaji, Ron Rapp, and Alan Press.

This book is dedicated to the love of my life, Holly, and our two darlings, Emily and Nathaniel. Without the love and support they provide, much would be left undone.

To God be the glory, honor, and praise.

chapter one

Introduction

"Society is always taken by surprise at any new example
of common sense."
                                              —Ralph Waldo Emerson

Systems engineering is more properly concerned with the *engineering of systems* than with merely definition of the requirements and architecture for such a system. It is also not merely concerned with analysis of performance and effectiveness at the "system level" of a system hierarchy. Systems engineering is really about common sense.

This book describes a well-structured process for developing systems, products, and services,[1] especially when these items are complex enough that conventional development techniques cannot deal with the development project's intricacies and uncertainties. We focus on defining the problem before designing the solution. We emphasize *user requirements* as being at least as important as the *technical requirements*. We recognize that requirements do not come merely from the customer and user. Many other *stakeholders* will supply requirements and constraints that will affect a product's development.

Even though the process is called systems engineering (SE), the SE activities described herein are not necessarily performed by members of an SE organization. In fact, the process is most often successfully implemented when multi-disciplinary teams are used.

This SE process is intended to be used on all development projects with appropriate tailoring as approved by management. Tailoring of

---

[1]Throughout this book, the term *product* is intended to *include* services. Services are merely a type of product and the term *product* will be used hereafter to mean both tangible products *and* services.

the process is expected and guidelines are given in Chapter 9. To be effective, tailoring should focus on reduction in scope of tasks, rather than elimination or drastic revision. The level of effort for each task should be adjusted according to the degree of acceptable development risk.

Any special terminology used in this book is defined where the term occurs or in the Glossary located in the back. Some of the special terms and concepts associated with Systems and Processes are defined and described in Chapters 2 and 3.

1.1 *Intended Users of this Book*

This book is intended for systems engineering managers, systems engineers, project[2] managers, design engineers, senior management, Process Champions,[3] process managers, and other supervisory personnel to provide familiarity with and understanding of the SE process, its use, and its value in the overall product development process.

This book may be used to support training of engineers in the SE process and as a reference guide for implementing the process. Non-SE Process Management Teams (PMTs) (e.g., software engineering, electrical engineering, etc.) will find this guide useful in coordinating their process tailoring and in defining interfaces. The generic SE process serves as the starting point for scaling and tailoring of the SE process for a specific project.

This book was written with the assumption that the reader has an understanding of the fundamentals[4] of systems engineering and its overall purpose. For example, the rationale is not given for why each task defined in this book is important. An understanding of the fundamentals of systems theory and the principles of good engineering practice is essential in properly implementing these tasks. Systems engineering training will normally be required before a complete understanding of the process can be realized.

---

[2]In this document, we make no distinction between "project" management and "program" management since one is a subset of the other. The term *project* is used to refer to the total organization that is responsible for developing a system or any of its products. The term *program* is used in this book for certain elements of a project such as the technical program, risk management program, verification program, and test program.

[3]The roles and responsibilities of a Process Champion are defined in Appendix D.

[4]This book focuses on the process itself. SE methods and principles are well described in many other books. See the Bibliography in this book for a listing of books that cover SE methods, principles, and other fundamental characteristics of systems engineering.

1.2 Definition of Systems Engineering

Systems Engineering basically consists of three elements:

a) *SE Management* plans, organizes, controls and directs the technical development of a system or its products.

b) *Requirements and Architecture Definition* defines the technical requirements based on the stakeholder requirements, defines a structure (or an architecture) for the system components, and allocates these requirements to the components of this architecture.

c) *System Integration and Verification* integrates the components of the architecture at each level of the architecture and verifies that the requirements for those components are met.

At any level in the architecture, a component can be passed to a development team for detail design of that component. The development of that component may use the SE Process if the nature of that component warrants. The SE Process should be applied to a component if any of the following are true:[5]

a) the component is complex

b) the component is not available off-the-shelf

c) the component requires special materials, services, techniques, or equipment for development, production, deployment, test, training, support, or disposal

d) the component cannot be designed entirely within one engineering discipline (e.g., mechanical design)

The actual nature of systems engineering is a complex subject fraught with controversy. Appendix A describes in much greater detail than in this Chapter "what systems engineering is" and "what systems engineering does."

Systems Engineering is the process that controls the technical system development effort with the goal of achieving an optimum balance of all system elements. It is a process that transforms a customer's needs into clearly defined system parameters and allocates and integrates those parameters to the various development disciplines needed to realize the system products and processes. Then, using the

---

[5]The actual criteria for a particular project should be defined in the engineering plan for the project.

analytical SE methodology, the process attempts to optimize the effectiveness and affordability of the system.[6]

"The Systems Engineering method recognizes each system is an integrated whole even though composed of diverse, specialized structures and subfunctions. It further recognizes that any system has a number of objectives and that the balance between them may differ widely from system to system. The methods seek to optimize the overall system functions according to the weighted objectives and to achieve maximum compatibility of its parts." [Chestnut, 1965]

Using the SE process described in this book fulfills two *fundamental purposes*:

* First, it makes sure we understand the question before designing the answer.

* Second, it coordinates, focuses, and balances the technical efforts of all involved throughout the development process.

"Understanding the question" is accomplished through various forms of disciplined analysis, focused on requirements and functions. Requirements are linked to design solutions through the allocation process in order to support traceability. Trade studies document the decision process.

The coordination and management aspects of SE include integration of disciplines as well as integration of system elements, risk management, technical progress monitoring, Design-To-Cost (DTC) control, technical reviews, documentation control, and overall planning and control of the technical effort.

1.3 SE and Concurrent Engineering

A Concurrent Engineering (CE) approach is inherent in the SE process as defined in this book. The SE process described in this guidebook involves a multi-disciplined, integrated, concurrent effort. SE solves complex problems, and requires extensive communication and interaction between and among the various engineering disciplines.

Concurrent engineering attempts to solve the problem of numerous, expensive design changes late in the development cycle. Figure 1-1 illustrates the results of using the CE concept [Clausing,

---

[6]We will use the term "system" throughout this document to mean a set of products and/or processes being developed (or being considered for development). A system could be a subsystem of a larger project, or it could be an aggregation of several other systems ("system of systems").

1991]. Design changes earlier in the development phases are much less expensive to correct, sometimes by several orders of magnitude.

The SE process recognizes that to manage a complex project, an integrated product development team approach is required (see Appendix B for a description of "The Multi-Disciplinary Team Concept"). Design, production, deployment, logistics, field support, and specialty engineers actively participate as members of multi-disciplinary teams in the execution of the SE process throughout the product development life cycle.

The number, size, and character of project teams will vary as the project progresses, with the lead shifting as a function of the tasks to be accomplished. The key to a successful team effort is for team members to be involved in the product definition and system optimization for the entire system and its life cycle rather than only for their part.

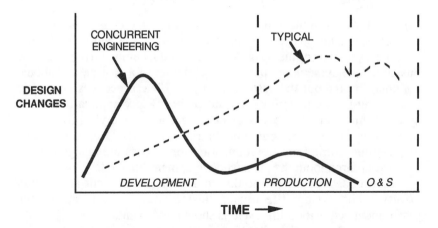

Figure 1-1 *Concurrent Engineering Approach versus Typical Approach*

1.4 SE Process Benefits

The concurrent involvement by design, production, deployment, logistics and specialty engineers results in front loading the design effort, but with the effect of reducing overall effort, costs, and technical changes in later stages of development. (See Figure 1-1.) Specific benefits that can be expected [LTV, 1992] are:

a) Shorter product development cycle time, reduced by as much as 60%

   b) Engineering change orders reduced by 50%

   c) Redesign and rework effort reduced by as much as 75%

   d) Manufacturing costs reduced by as much as 40%.

Therefore, this leads to the following general benefits:

   a) Reduced non-recurring costs

   b) Reduced life cycle costs

   c) A more competitive product

   d) A process that is compliant with the expectations and requirements of external customers.

1.5 Applications of the SE Process

This generic systems engineering process is consistent with the industry standards on systems engineering—EIA 632 and IEEE 1220.

   The systems engineering process should begin during the new business development and pre-proposal phases of a project, and should continue throughout the system life cycle. The process is critical in responding to a request for proposal (RFP). Early application of the systems engineering process ensures that a company's proposal is responsive to all the customer's requirements. Initial systems engineering efforts form the foundation for successful management of the project throughout the product development life cycle.

   This process can, and should, be used in government acquisition efforts. The system life cycle, "lust to dust,"[7] in terms of the government acquisition life cycle is shown in Appendix C.

   Systems Engineering should be applied throughout the system life cycle as a comprehensive, iterative technical process to:

---

[7]The expression "lust to dust" has recently been replacing the traditional phrase "cradle to grave" due to the rising concerns with environmental problems in disposing of a system's waste by-products (and sometimes the whole system itself). Think about it. How much does it cost to maintain a nuclear waste depository for hundreds if not thousands of years. Was this cost considered in the original cost-effectiveness analysis for the nuclear plant?

Also, there is now more awareness that the SE process needs to be applied prior to a concept's "birth" (i.e., the "cradle"). Often at birth, the major constraints have already been decided and there may be little room to synthesize better solutions. The analysis of requirements and technology options needs to occur before the birth, as far back as the first desire for a solution to a problem (i.e., the point in time when there is a "lust" for a solution to a problem at hand).

- Translate the customer's need into a configured product or system meeting that need through a systematic, concurrent approach to integrated design of the product and its related manufacturing, test and support processes.

- Integrate the technical inputs of the entire development community and all technical disciplines (including the concurrent engineering of manufacturing, logistics and test) into a coordinated effort that meets established project cost, schedule and performance objectives.

- Ensure the definition and compatibility of all functional and physical interfaces (internal and external) and ensure that system definition and design reflect the requirements of all system elements (hardware, software, facilities, people, data, services, materials and techniques).

- Identify, define and classify technical risks, develop risk abatement approaches, and reduce technical risk through analysis and early test and demonstration of system elements.

1.6 Key Elements of Systems Engineering

There are six, no seven, key elements of a successful systems engineering effort:

- **Systems Engineering Management Plan (SEMP).** The SEMP is a comprehensive document that describes how the *fully integrated engineering effort* will be managed and conducted. The SEMP is described in Section 4.4.

- **Systems Engineering Master Schedule (SEMS).** The SEMS is a compilation of the key development tasks and milestones with their associated accomplishment criteria. The SEMS is *event-based* and is described in Section 4.5.

- **Systems Engineering Detailed Schedule (SEDS).** The SEDS is a detailed, task-oriented schedule of the work efforts required to support the events and tasks identified in the SEMS. The SEDS is *calendar-based* and is described in Section 4.5.

- **Work Breakdown Structure (WBS).** The WBS is a product-oriented family tree composed of hardware, software, facilities, data and services which result from the SE efforts during the development and production of the system and its components, and which completely defines the project. It displays and defines the products and processes to be developed or produced, and relates the elements of work to be accomplished to each other and to the end product. The WBS provides structure for guiding multi-disciplinary team assignment and cost tracking and control. The WBS is expanded and refined during the Synthesis tasks described in Section 7.3. Assessment of project progress is performed with respect to the WBS during the SE Control and Integration tasks described in Section 6.2.

- **Requirements.** Requirements are characteristics that identify the accomplishment levels needed to achieve specific objectives for a given set of conditions. A requirement can be either a threshold or an objective.[8] Primary (driving) requirements are defined during the Requirements Analysis tasks described in Section 7.1. Derived requirements are defined and allocated to functions and system components during the Functional Analysis/Allocation tasks and the Synthesis tasks described in Sections 7.2 and 7.3, respectively. Contractually binding technical requirements are stated in approved specifications. Specifications are described in Section 7.5 and in the Glossary. Design-To-Cost (DTC) objectives should be treated as requirements and may be documented in a Cost Breakdown Structure (CBS). (See [Blanchard, 1990] for details on the CBS.)

- **Technical Performance Measurement (TPM).** Technical Performance Measurement (TPM) is a technique for assessing progress toward meeting a technical performance requirement or objective. The expected performance is determined *a priori* for different points in time for the project. The actual (or predicted) performance is compared against the preplanned expected performance. Any parameter that has not achieved its planned level for that point in the project is a candidate for risk mitigation or for a proposed change in requirement either through reallocation of that performance or a change in the driving requirement.

- **Technical Reviews and Audits.** Reviews and audits are activities by which the technical progress of a project is assessed relative to its technical and contractual requirements. They are conducted at logical transition points in the development effort to reduce risk by identifying and correcting problems or issues resulting from the work completed before the project is disrupted or delayed. Technical reviews and audits are described in Section 6.2 (Task 208) and Appendix F.

---

[8]*Thresholds* are basically an absolute limit of something that *must* be achieved. An *objective* has some degree of usefulness or desirability, but is not necessarily mandatory.

1.7 Key Issues of Systems Engineering

The SE Process, if followed and tailored properly, will contribute to higher quality products, shorter development cycles, and lower cost products. However, for the SE process to be successful, there are some issues that must be highlighted to ensure they are addressed. The key issues are described below.

1.7.1 Need For a SEMP

The SEMP should be derived from and developed in parallel with the Project Management Plan and must cover all the technical activities associated with performance of the engineering activities for a project. This document establishes the technical program organization, direction and control mechanisms for the project to meet its cost, schedule and performance objectives.

The SEMP defines and describes the systems engineering management, the tailored SE process and how the technical disciplines will be integrated for the life of a project. Since this is the primary engineering planning document, every project, regardless of size, should develop a SEMP. The SEMP should be tailored to project size, complexity and cover all development phases. As the system development progresses, the SEMP must be regularly updated to reflect any new project requirements and changes in activities.

1.7.2 Need For a Fully Integrated SEMS and SEDS

The SEMS is a list of the *key events* in a project along with their associated *completion criteria*. The SEDS is a detailed schedule of the *technical tasks* required to complete the events and major tasks identified in the SEMS. The SEDS will outline detailed tasks, their dependencies, and start/end dates for each SEMS event and task. The SEDS will be regularly updated to reflect project requirements and the relationship between the SEMS and SEDS should be maintained throughout the project.

1.7.3 Establish Multi-Disciplinary Teams

A key ingredient to a project's success is the establishment of multi-disciplinary teams to perform Concurrent Engineering (CE). (See Chapter 11 and Appendix B for a discussion of the multi-disciplinary team concept.) The emphasis on the integration of product and process has placed new demands on groups whose work is interrelated to communicate effectively. The fundamental advantage of these teams is that many issues can be resolved quickly through the direct communications among team members, resulting in shorter cycles and improved product and process performance.

Team membership will vary in each project phase and task. A team is most effective when its membership is drawn from representatives of systems engineering, electrical design, mechanical design, software development, manufacturing, quality, specialty engineering, purchasing, maintenance and suppliers. Using the SEMP, SEDS, WBS and the SEMS, management can establish the proper team at the proper time to address the correct problem or issue.

1.7.4 Establish Design-to-Cost (DTC) Objectives

A DTC management plan is a key element in a strong cost containment effort. The plan must be developed and implemented early in the project to achieve desired results. The success of the plan is a direct function of how well the specific cost goals are allocated to the specific tasks in the WBS. DTC goals can more easily be met if they are treated as a requirement of equal importance, in general, to any technical requirement.

1.7.5 Effective Use of Reviews

Reviews, both contract-mandated and internal, are an effective technique for improving quality, containing costs and staying on schedule. Reviews should occur throughout the product life-cycle. However they must be planned early in the project and must assess technical maturity. Reviews should be conducted when the planned entrance criteria have been met and not before.

Management must support the review process by providing time and resources, without sacrificing either for the schedule. Effective reviews must be tailored to verify the level of design maturity. Reviews are effective when experienced individuals (domain experts) participate in these reviews to ensure that problems are highlighted and resolved.

1.7.6 Integrate the Responsibilities of Project Management and Systems Engineering

The SE Management tasks defined in this book are *not necessarily performed by members of an SE organization*. These tasks could be performed by Project Managers, SE Managers, Integrated Product Team (IPT) leaders, or IPT members. Responsibilities must be clearly defined and allocated to avoid duplication of efforts between SE and

Project Management (PM). Responsibilities need to be clearly mapped from the WBS to the Organizational Breakdown Structure (OBS) using a Responsibility Assignment Matrix (RAM).

1.7.7 Effective Use of Simulation Tools

Simulations can improve the decision process by exploring several alternative designs and enhance the trade-off studies between the designs. Using simulations can prove, in relatively short time, that the correct product or system is being designed to meet the performance criteria. Early prototyping, through simulations, can reduce system development cycles, minimize risks, improve reliability and improve performance.

1.7.8 Establish Common Tools and Databases

Having common databases that are accessible by all technical disciplines does not guarantee effective communications. However, it does ensure that all disciplines are using the same information to make their decisions and everyone can see the results instantly.

With common databases, redundant information can be eliminated, access control is enhanced and data security issues are minimized. Also, product and process information can be more effectively developed concurrently, reducing the development cycle time (i.e., time to market). Successful implementation of tools (CAE, CASE, CIM, CAM, etc.) and integrated databases requires:

- Planning both the short- and long-term strategies;
- Re-engineering the process to take advantage of the technology available;
- Integrating the various tools and databases, and, where necessary, developing the interfaces between tools and databases for compatibility;
- Providing the necessary training;
- Maintaining the tools and databases; and
- Evaluating the process with metrics to identify discrepancies in requirements or implementation.

Some of the benefits that can be expected are improved productivity and quality, and faster time to market.

1.7.9 Follow a Systematic Approach to Systems Development

A systems engineer usually starts with the customer requirements and system architecture, decomposes the system into the subsystem and lower-level subsystems or components, and then uses systems analysis to develop the requirements, using a verification/test program to ensure the requirements are met and testable. Using a top-down, structured approach will increase the chances that the ultimate end user of the system or its products will be satisfied.

1.8 SE Process Tailoring

The goal of defining the systems engineering process in this book is to provide a structured, tailorable framework for all systems engineering efforts in an organization, regardless of size, complexity, and customer requirements. Specific requirements and characteristics of individual projects will be reflected by tailoring of the process. However, the execution of the underlying standard process will yield major, measurable improvements in the development, production and deployment of an organization's products.

Each project should have a Process Champion who will be responsible for tailoring the SE process for the project, assisting in documentation of the tailored process in a SEMP, providing training for project members, and collecting process metrics. The roles and responsibilities of the Process Champion are further described in Appendix D.

1.9 Key Questions of Systems Engineering

The essence of systems engineering can perhaps be illustrated by listing the "key questions" that might be asked when engineering a system. These questions can be categorized as shown in Figure 1-2 by notional "phases."

One of the crucial aspects of systems engineering is the ability to "separate concerns." This is facilitated by dividing the overall engineering activity into the relevant phases and "peeling back the onion" one layer at a time. It is often the case that all of the phases shown in Figure 1-2 are repeated for each layer of the onion. This is called *recursion*, another essential attribute of systems engineering.

Need

⇩

- What needs are we trying to fill?
- What is wrong with the current situation?
- Is the need clearly articulated?

Operations Concept

⇩

- Who are the intended users?
- How will they use our products?
- How is this different from the present?

Functional Requirements

⇩

- What specific service will we provide?
- To what level of detail?
- Are element interfaces well defined?

System Architecture

⇩

- What is the overall plan of attack?
- What elements make up the overall approach?
- Are these complete, logical, and consistent?

Allocated Requirements

⇩

- Which elements address which requirements?
- Is the allocation appropriate?
- Are there any unnecessary requirements?

Detailed Design

⇩

- Are the details correct?
- Do they meet the requirements?
- Are the interfaces satisfied?

Implementation

⇩

- Will the solution be satisfactory in terms of cost and schedule?
- Can we reuse existing pieces?

Test

- What is our evidence of success
- Will the customer be happy?
- Will the users' needs be met?

Figure 1-2   *Key Questions of Systems Engineering*

1.10 The Ideal Systems Engineer

What are the ideal traits of a systems engineer? Arthur D. Hall in his classic book, *A Methodology for Systems Engineering*, defines the following traits for an "ideal systems engineer" [Kasser, 1995]:

*   *An ability to see the big picture*—the most important trait; it means that the systems engineer is not concerned primarily with the devices that make up a system, but with the concept of the system as a whole—its relations and its behavior in the given environment

*   *Objectivity*—the ability to appraise with complete objectivity

*   *Creativity*—a vital part of the SDLC (System Development Life Cycle)

*   *Human relations*—not just the ability to get along with people, but the positive attributes of leadership, tact, diplomacy, and helpful concern so essential in effective teamwork

*   *A broker of information*—the gift of expression—oral, written or sometimes graphic

*   *Education*—graduate training in the relevant field of interest (application) as well as courses in probability, statistics, philosophy, economics, psychology, and language (logic and business management should also be included)

*   *Experience*—experience in research, development, systems engineering and operations

References

Blanchard, B., and W. Fabrycky, *Systems Engineering and Analysis*, 2nd ed. Prentice Hall, 1990.

Chestnut, H., *System Engineering Tools*. John Wiley, 1965.

Clausing, D., *Study of Japanese Auto Manufacturers vs. Ford Motor Company*. MIT Press, 1991.

Hall, Arthur D., *A Methodology for Systems Engineering*. Van Nostrand, 1962.

Kasser, Joe, *Applying Total Quality Management to Systems Engineering.* Artech House, 1995.

LTV, *Concurrent Engineering,* a report by LTV Aerospace and Defense Company at US Air Force/US Army Concurrent Engineering Applications Symposium, Ft. Walton Beach, FL, March 31, 1992. Results are based on recent surveys comparing actuals to forecasts or norms for the type of development effort involved.

chapter two

Systems concepts

"Toto, I have a feeling we are not in Kansas anymore."
—Dorothy, in *The Wizard of Oz*

Some of the basic systems concepts essential to effective implementation of the systems engineering process are described in this Chapter.[9]

2.1 Systems Terminology

2.1.1 Definition of System

The term *system* is used to mean a set of integrated *end products* and their *enabling products*.[10] The end products and enabling products of a system are composed of one or more of the following: hardware, software, personnel, facilities, data, materials, services, and techniques.

2.1.2 Definition of Architecture

Architecture can be defined as: "The highest-level concept of a system in its environment. An architectural description is a model—document, product or other artifact—to communicate and record a system's architecture. An architectural description conveys a set of

---

[9]This Chapter (except for Section 2.5) is based on the original work of Richard Harwell, Dr. J. G. Lake, Dr. John Velman and the author in preparation of the updated EIA standard on systems engineering (EIA 632). All rights are reserved. Used with permission.

[10]The terms *end products* and *enabling products* will be defined in Section 2.2.1.

views each of which depicts the system by describing domain concerns." [SESC, 1996]

So, an architecture deals with a system structure, its operational interfaces, profiles of use, and how the elements of a system interact with each other. An architecture can also be defined in terms of scenarios along with expected behavior for each scenario; states, modes and configurations of the system elements; and basic system purpose or mission.

2.1.3 Definition of Requirement

A *requirement* is: (1) A characteristic that identifies the accomplishment levels needed to achieve specific objectives under a given set of conditions. (2) A binding statement in a document or in a contract.

Requirements are of the following three basic types: *functional, performance,* and *constraint.* Examples of constraint types are: legislative, legal, political, policy, procedural, moral, technology, interface. Requirement statements are often hybrid combinations of the basic types. For example, a requirement statement might define the function to be performed along with the required performance.

For each requirement, the applicable conditions must be defined. Examples of condition types are: environmental, state, mode, and configuration. Mandatory requirements (thresholds) are traditionally expressed using the word "shall." Recommendations (non-mandatory requirements, otherwise known as objectives)[11] are traditionally expressed using the word "should."

A *functional requirement* specifies: (1) The necessary task, action, or activity that must be accomplished, or (2) *what* the system or one of its products must do. A *performance requirement* specifies *how well* the system or one of its products must perform a function along with the conditions under which the function is performed.

A *derived requirement* is: (1) A requirement that is further refined from a primary source requirement or from a higher level derived requirement. (2) A requirement that results from choosing a specific implementation for a system element.

Further information on the nature of requirements is given in Section 2.5.

---

[11]Some of you may be wondering how you can have a "non-mandatory" requirement. This is one of the confusions with regard to what a requirement is. If a requirement is anything that is needed, then it is very likely that it will be impossible for any one solution to meet all of someone's needs. Trade-offs will have to be made, compromises reached. That is why it is essential when "gathering" requirements that priorities (or relative value) for each are established. These priorities will ensure that the proper trade-offs are made.

2.1.4 Distinction Between Stakeholder, Customer, and User

The source of requirements is not only from *customers*, but also from other *stakeholders* that have a need or expectation with respect to system products or outcomes of their development and use.

Examples of *stakeholders* are acquirer, user, customer, manufacturer, installer, tester, maintainer, executive manager, and project manager. The corporation or agency as a whole and the general public are also stakeholders for a system.

A *customer* is either an individual or organization that (1) commissions the engineering of a system, or (2) is a prospective purchaser of an end product. A *user* is either an individual or organization that uses the end products of a system.

2.1.5 Distinction Between Stakeholder Requirements and Technical Requirements

Stakeholder requirements are often stated in non-technical terms (e.g., needs, wants, desires, and expectations) and are not normally adequate for design purposes. Also, the stakeholder requirements may not be verifiable using normal technical verification techniques. However, they do provide the measures of effectiveness by which delivered end products will be judged by that stakeholder.

Also, stakeholder requirements already stated in technical terms from higher layers often need to be translated into technical requirements appropriate for *that layer of development.*

Technical requirements are derived from stakeholder requirements and are stated in clear, unambiguous, and measurable technical terms. Technical requirements are verifiable and directly traceable to stakeholder requirements and define the technical problem to be solved.

2.1.6 Distinction Between Requirements and Design

A *requirement* deals with what is acceptable to a stakeholder—both in terms of thresholds and objectives. *Design* deals with what is achievable through the application of technology. Design is literally a concept of the mind, an invention, created to meet a need. A design captures the intention of the designer. Design features are part of a product for various reasons:

a) conscious decision of a designer

b) unconscious decision of a designer

   c)   decision or action of a builder, tester, installer, user, etc.

Since design features can be added at any point in the process of getting a product into service, it is important to ensure that these added features do not violate any stakeholder requirements. Extra features also can add cost, decrease safety or quality, reduce product life, etc.

2.1.7 Distinction Between Requirements and Specifications

Requirements document the needs of various stakeholders. *Specifications* are documents[12] that contain the requirements that have been *agreed upon* with a particular set of stakeholders. Usually before the specification agreement is established, the downstream receivers of the specification are solicited for comments on the feasibility of achieving the specified requirements.

A common mistake is to treat all requirements as equally important. Even worse is assuming the wrong priority for a requirement. Specifications have rarely contained priorities assigned to each requirement, but it would go a long way toward getting lower cost and faster delivery if the *negotiable* elements are spelled out clearly.

2.1.8 Distinction Between Verification and Validation

Validation and *verification* are tasks associated with activities within the systems engineering process. *Verification* ensures that the selected solution meets its specified technical requirements, and properly integrates with interfacing products.

Validation ensures that the requirements are consistent and complete with respect to higher level requirements. Validation ensures that "you are working the *right problem,*" whereas verification ensures that "you have solved the *problem right.*"

Validation will often involve going back directly to the users to have them perform some sort of acceptance test under their own local conditions. A common mistake is to wait until the system has been entirely integrated and tested (the design is qualified), prior to performing any sort of validation. Validation should occur as early as possible in the product life cycle.

---

[12]Often, the term *specification* is used to mean the actual statement of a required characteristic. In this book, the term *specification* is used only to refer to the *document* that contains these required characteristics.

When the first set of technical requirements is derived from the user requirements and other stakeholder requirements, these technical requirements should be validated against these *source requirements*. Full requirements traceability is essential to facilitate early validation of requirements. Requirements flowdown is discussed further in Section 2.5.3.2.

2.2 System Hierarchy

Critical requirements for a system are often missed because the full life cycle of a system is not well understood or the enabling products for the system are not well defined. Enabling products are those products that provide the means for getting end products into service, keeping them in service, and for properly retiring those end products from service.

This Section describes a system life cycle, defines enabling product types, and gives examples of enabling products for a system. The utility of developing a full definition of the system life cycle and its enabling products is described. A mapping between enabling product types and the system life cycle is also described. Full consideration of the system life cycle and its enabling products will ensure greater coverage of the requirements needed for the engineering of systems.

The concept of a "building block" is defined below. The purpose of the building block is to serve as a "framework" for application of the systems engineering process. The systems engineering process is *applied at least once* to each building block in the total system.

2.2.1 System Building Block

The IEEE 1220 standard on systems engineering defines a "basic building block of a system" as:

> *the system, its related product(s), the life cycle processes required to support the products and customers for the product(s), and the subsystems that make up the product(s).*

The 1220 standard uses this basic building block (see Figure 2-1) to define a larger structure called a system breakdown structure (SBS). A simpler form of the building block has been developed and is fully described in "Anatomy of the Engineering of a System" [Harwell 1996]. This Section will focus on the relationship between the "enabling products" shown in the building block and the life cycle of the system.

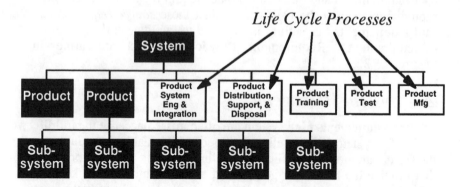

Figure 2-1   *Basic Building Block from the IEEE 1220 Standard*

This new approach deals with five separate entities: (1) a building block *system*, (2) end product *subsystems*, (3) *end products*, (4) *enabling subsystems*, and (5) *enabling products*. Figure 2-2 shows the relationship between the first four of these items (similar to the 1220 basic building block above).

There are enabling products associated with each enabling subsystem. For a project developing a system or product, there is a hierarchy of these building blocks that make up the total system. Each "tier" of this hierarchy is called a "development layer." This new approach will help to ensure that all the relevant requirements for a system development are properly considered.

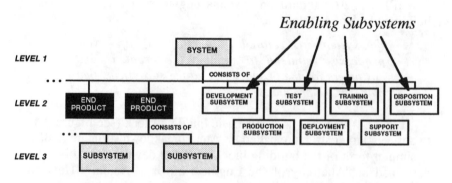

Figure 2-2   *A Different View of the Basic Building Block*

2.2.1.1 Recursive Application of the Process

The systems engineering process is applied recursively on each building block at each development layer. The development layers correspond to the physical "tiers" of the system itself. The project[13] and its subprojects each have a separate *building block system*. Each building block system requires one or more *end products*. (The end product is the thing that "ends up" in the hands of the user.) Each end product may require two or more *end product subsystems*. If an end product requires no development, then there is no need to define subsystems for that item. The subsystems are treated as building block systems at the next lower layer of development (see Section 2.2.3 for discussion of development layers).

2.2.1.2 Enabling Subsystems

The white boxes in Figure 2-2 are called *enabling subsystems*. The enabling subsystems represent the system life cycle considerations in development of the end products. Each enabling subsystem has enabling products associated with it. For example, test equipment is an enabling product for the Test Subsystem.

2.2.1.3 Enabling Products

Each building block system requires *enabling products* that provide the means for getting the end products (for that development layer) into service, keeping them in service (e.g., end products properly maintained and operators properly trained), and for properly retiring those products from service. The enabling products perform enabling functions on the end products, or on related components of the system, such as personnel, to ensure proper operation of the system and its components.

The enabling products for one development layer might have no relation to the enabling products for another development layer. For example, the production facility for integrating the subsystems at the system level may be different from the production facility used to produce any of the subsystems. The concepts above are discussed in more detail in [Harwell 1996].

---

[13] A project is an organizational structure within a business organization or other agency used to develop and/or produce a system or product. A subproject is subordinate to this project and may be performed by a different organization, company or agency. Often the subproject is itself called a project by those who are in that organization and it too can have lower level subprojects. See Section 2.3.1 for discussion on the project interfaces.

2.2.2 Basic Product Types

A system is composed of various components. The system itself is a component, and you could also say that each component is a system. A single component consists of one or more of the basic product types shown in Figure 2-3. Notice that a system product is not merely hardware or software. There are many other types of products that can perform the necessary functions that meet stakeholder needs.

A product is any one of the system components that needs to be produced or acquired. Some components can be acquired (i.e., procured) as-is without need for fabrication or modification.

2.2.2.1 Product Taxonomy

These basic product types are not necessarily mutually exclusive. For example, some would consider that facilities *contain* hardware and people. Others would consider facilities to be separate from hardware and people. Some say that services are really techniques. The important thing to remember is to include all of the right components in the system being developed. Having a "taxonomy" of product types can serve as a checklist to ensure that all bases are covered.

Also, there are certain to be other basic product types, some only relevant to certain industries. The important thing to remember is that required behavior for a system should not be allocated merely to the hardware and software elements.

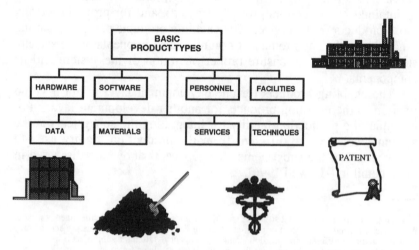

Figure 2-3 *Basic Product Types*

2.2.2.2 Product Type Examples

Examples of each product type are shown below.

Hardware	Computer processor unit, pencil, radar set, paper, satellite bus structure, telephone, diesel engine
Software	PC operating system, firmware, satellite control algorithm, robot control code, telephone switching software
Personnel	Astronaut, PC operator, clerk, business executive, Laika (the cosmonaut dog), bus driver, cashier
Facilities	Space rocket launch pad, warehouse building, shipping docks, airport runway, railroad tracks, conference room, traffic tunnel, bridge, local area network cables
Data	Personnel records, satellite telemetry data, command and control instructions, customer satisfaction scores
Materials	Graphite composite, gold, concrete, stone, fiberglass, radar absorption material, cladded metals
Services	Overnight delivery, long distance telephone, battlefield surveillance, pizza delivery, lawn maintenance
Techniques	Soldering, trouble ticket response process, change notice handling, telephone answering protocol, project scheduling

Materials could be thought of as your basic raw materials, like steel, or as complex materials, like cladded metals, graphite composites, or building aggregate material. Personnel are not normally thought of as a "product," but that depends on which system you are working with. The NASA space program "system" certainly produces astronauts; you cannot just go out and hire one! Services are sometimes thought to be the functions that your system performs. Services usually require special hardware and software to gain "access" to that service.

Services could also be functions that an external system performs that are not under your control such as the friendly IRS. But, seriously, sometimes the most cost-effective solution is to allocate a function to someone else's system. This is one of the most overlooked methods for developing the most affordable and quick-to-market system.

2.2.2.3 Full Definition of a Product

A *full product definition* includes requirements, concepts, structures, and terminology for that product:

a) *requirements* for what the component does and how well and under what conditions and any applicable constraints

b) *concepts* for interaction with the environment and external components during all life cycle activities of development, production, test, distribution, installation, operation, support, and disposition (examples are operations concept and maintenance concept)

c) *structures* for how the component relates to itself and to other components (this is sometimes called architecture), includes superstructure for structure outside the component and substructure for what the component consists of. All relational attributes are included here such as interfaces, ownership, configurations, etc.

d) *terminology* for what the elements of the component are called plus terms for the unique items in the requirements, concepts, or structures.

2.2.3 Development Layers

A single building block will rarely define the complete solution to a complex problem more typical of today's projects. If an end product subsystem requires further development, it will have its own subordinate building block. Once the descriptions of end products of the initial building block are completed, and preliminary descriptions of the end product subsystems are defined, the development of the next lower layer of building blocks can be initiated.

The building blocks are connected to form a hierarchy as shown in Figure 2-4. This layered approach in the aggregation of building blocks is continued until:

(1) the subsystems or end products of a building block can be *manufactured,*

(2) the requirements for a subsystem or end product can be satisfied by an *existing product,* or

(3) the subsystems or end products can be *procured* from a supplier.

The particular building block structure will vary for each solution.

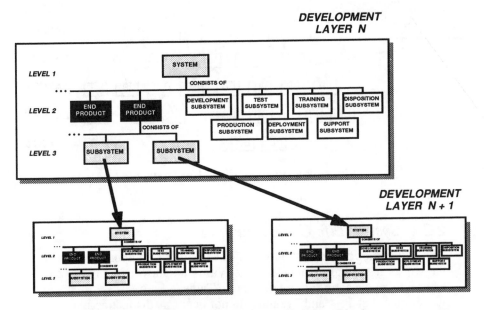

Figure 2-4 *Relationship Between Building Blocks*

Figure 2-5 shows how a building block hierarchy for an entire project might look. The building block has reached the "bottom" when the end products or their subsystems can be either procured from vendors or the design of these items requires "no development" (i.e., all enabling products for that end product already exist and are all compatible with each other and with the total solution).

Notice that a building block can reach bottom at any development layer. Building blocks will continue to be defined as long as further development is required. In the case shown below, the systems engineering process is applied at least 27 times, once for each building block.

The process will be applied further for any *enabling product* that needs its own development. For example, if a particular test equipment has not been designed yet, it will need its own development, production, test, deployment, etc. (i.e., its own enabling subsystems). In a sense the enabling product (i.e., test equipment, in this example) becomes an end product for a different end user—the test engineer. This is the *recursive nature* of systems engineering.

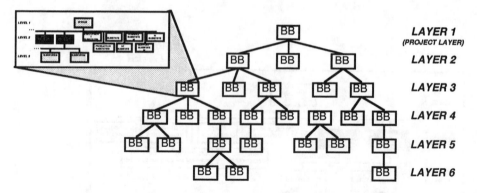

Figure 2-5 *Building Block Hierarchy*

2.2.4 System Life Cycle Activities

The life cycle for the system or any of its major components can be shown as a typical sequence of activities as shown in Figure 2-6. These life cycle activities are defined in Table 2-1.

Notice that Test and Training do not occur as distinct activities in this diagram. Training is typically part of the Support activity but can also occur at any point in time during the system life cycle. For example, the production personnel may be trained in the operation of the equipment they will produce.

Test typically occurs everywhere. The products will be tested during Development by engineers, during Production by production personnel, perhaps even during Operations by the operators themselves.

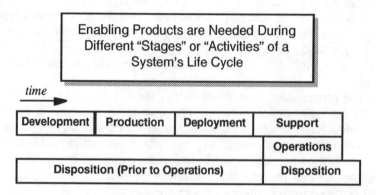

Figure 2-6 *Life Cycle Activities for a Single System or Component*

Table 2-1   *Definitions of Life Cycle Activities*

Development	Analyzing the problem and synthesizing the solution. Building models to check compliance to requirements and to validate design assumptions. Ends when design is fully qualified, all requirements have been complied with or appropriate deviations have been approved. Webster's defines as: "to work out the possibilities of ...; to make available or usable."
Production	Manufacture of production units for use by users.
Deployment	Getting the production units to the warehouse, or to the retail store, ready for procurement or delivery and installation. Gets product to the "point of sale." Includes getting production units delivered, installed and ready for operational use.
Support	Includes preventive and corrective maintenance throughout operational life. Includes training the users, maintainers, disposers, administrators, etc. Includes supply of consumables (fuel, chemicals, paper) and of spare parts. Includes administration of products in the field (e.g. changing your phone service to include "call waiting" is an administration activity). It is an administrative task to determine when maintenance is needed and to schedule the maintenance tasks.
Operations	Using of the end products.
Disposition	Prior to operation, the activities required to dispose of waste by-products, scrap, obsolete equipment, etc. During operations, includes activity required to know status of operational units and to dispose of waste products during its operational life (e.g. consumables, energy devices, failed components, etc.) and dispose of product/system at end of its useful life. Could also involve sending back a product for refurbishment or long-term storage. Refurbishment sometimes requires new development to replace obsolete parts or to upgrade to newer technology. The refurbished article is essentially starting a new life, hence, a repeat of the life cycle.

NOTES—

(1) The 1220 standard uses the term "Distribution" instead of "Deployment" to include installation, test, checkout and train. This is contrary to the normal use of that term which says that distribution gets the product to the "point of sale." This is solved by defining Deployment to include getting the product from the point of sale to the operational state via delivery, installation and checkout.

(2) The 1220 standard defines "Test" as a separate "life cycle process" (see Figure 2-1). However, Test is not just the activity that happens to a production unit. Test is not shown as a separate life cycle activity since Test-type enabling products are used throughout the life cycle.

(3) Disposition fixes the problem with Disposal not including the other activities required to dispose of products. You have to know the status of the products in order to properly dispose of the product or waste by-products—its location, its state of usage (e.g. hours in service), its state of wear, its efficiency (efficiency usually drops with product age), etc.

(4) Administration is one of the tasks described above under the Support activity. In the telecommunications world, they talk of OA&M. The A is for Administration because there is a lot of administration required for the telephone networks, switches, and services. The rest of the world usually only talks of O&M.

2.2.5 System Enabling Products

It is probably not possible to define all the required enabling product types for all industry sectors. The most we can probably do in this Section is to give examples and to say that "The Developer shall define the enabling product types for their industry sector," and "The Developer shall identify and define the enabling products for each development layer of the SBS."

Table 2-2 shows a mapping between the enabling product types and the life cycle activities. Notice that Deployment uses enabling product types of Distribution, Installation, Checkout and others. Support uses such types as Administration, Maintenance, Refurbishment, among others.

There are certainly other enabling product types that are not shown below. However, it is probably true that the LC activities defined herein are true and complete for all possible systems and products.

(Of course, every industry sector will have different names for these enabling subsystems and will have different mapping of tasks to these activities. It is left to the reader to determine how their own industry defines a system life cycle. The important thing is to ensure that the "full" life cycle for your system is defined— "lust to dust" as they say—regardless of what terminology you use.)

2.2.5.1 Examples of Enabling Products

Table 2-3 shows examples of enabling products for each type. (Remember—the definition of a "product" includes hardware, software, personnel, facilities, data, materials, services and techniques.) Each enabling product required for a particular system development needs to be either procured, developed, or defined.

Personnel. Where personnel are shown below as an example enabling product, the persons are not actually developed, but the required or existing characteristics of those people need to be defined. The characteristics needing definition are: Knowledge, Skills, Abilities, and Motivation (known as the KSAMs in organizational behavior parlance). In a sense, the purpose of the Training activity is to "develop" the particular KSAMs in the trainee that are deficient.

Multiple uses. Many of these products can have multiple uses throughout the life cycle. For example, an engineering model built to validate design assumptions could be used later to provide training to users, serve as a mass mockup during integration, serve as a shop aid during production, and be used as a trade show model during marketing activities. During the development of the System Breakdown Structure (SBS), the multiple uses of enabling products should be defined and possible conflicts identified and resolved.

The enabling subsystems of a building block provide the means by which the life cycle activities are implemented for the (operations) end product. Each enabling subsystem may have its own enabling products that provide the means for getting end products into use, maintaining end product readiness, and dispositioning of end products after use.

Table 2-2   ***Mapping of Enabling Products to the Life Cycle Activities***

Enabling Product Types	Develop-ment	Produc-tion	Deploy-ment	Support	Opera-tions	Dispo-sition
Development	X					
Manufacturing	X	X		X		
Integration	X	X	X			X
Test	X	X	X	X		X
Distribution			X			
Installation			X	X		X
Checkout			X	X		
Training		X	X	X	X	X
Operations					X	
Administration			X	X		X
Maintenance				X		X
Refurbishment		X	X	X		X
Disposal	X	X	X	X	X	X
Marketing	X	X		X		X
et cetera						

NOTES—
(1) Not all of these enabling products are relevant in all industry sectors nor technology domains. For example, for an industry that only sells services (such as your local telephone company or your Internet service provider), there is no need to dispose of anything related to a service.
(2) There may be other enabling products not shown here depending on the industry sector.

Table 2-3 **Examples Of Enabling Products**

Enabling Subsystem	Example Enabling Products
Development	Development plans and schedules, engineering policies and procedures, integration plans and procedures, shared database, automated tools, analytical models, physical models, cables and other interface structures (not included in end products or other enabling subsystems)
Production	Production plans and schedules, manufacturing policies and procedures, manufacturing facilities, jigs, special tools and equipment, production processes and materials, production and assembly manuals, measuring devices, personnel certification criteria
Test	Test plans and schedules, test policies and procedures, test models, mass/volume mockups, special tools and test equipment, test stands, special test facilities and sites, measuring devices, simulation or analytical models, demonstration and scale test models, inspection procedures, personnel certification criteria
Deployment	Deployment plans and schedules, deployment policies and procedures, mass/volume mockups, packaging materials, special storage facilities and sites, special handling equipment, special transportation equipment and facilities, installation procedures, installation brackets and cables, special transportation equipment, deployment instructions, ship alteration drawings, site layout drawings
Training	Training plans and schedules, training policies and procedures, simulators, training models, training courses and materials, special training facilities, operator/maintainer certification criteria
Support	Support plans and schedules, support policies and procedures, special tools and repair equipment, maintenance assist modules, specials services (e.g., telephone hotline and customer access lines), special support facilities and handling equipment, maintenance manuals, maintenance records system, special diagnostic equipment (not an integral part of the end product)
Disposition	Disposition plans and schedules, disposition policies and procedures, refurbishment facilities and equipment, special disposal facilities and sites, special equipment for disposal of spent end products

2.2.5.2 *Importance of Enabling Products*

To ensure that system development is proceeding without undue risk, the full component definition for all relevant enabling products and the enabling subsystems for the system, its end products and enabling products must be well defined and understood.

The full component definition consists of requirements, concepts, structures, and terminology for that product. It is important to identify all relevant enabling products for the end products. Often adequate consideration is not given to these items and this may result in cost overruns, schedule slips, and project cancellations.

The concepts described in this Section can help to ensure that "due process" is given to the full life cycle considerations for a system. This should result in more satisfaction for end users and other stakeholders of the system.

2.2.6 System Integration

It has been said that systems engineering is concerned with the *integration* of system elements. This is certainly true, but there are some critical things left unsaid. What *system* and what *elements* are to be integrated?

It turns out that there are several systems of concern in any application of the systems engineering process. First, there is the system that ends up in the hands of the user. This is sometimes called the end product. There is also the system that manufactures the end product. There are also systems for verification (or test), deployment, installation, training, maintenance, administration, and disposition.

There are also all the non-product related elements such as process, methods, tools, and environment. Also, how about technology and people? Hence, there are many "elements" that need to be integrated. This Section describes the things that need to be integrated and some of the means by which *full integration* can be ensured.

2.2.6.1 Definition of System Integration

"System Integration" is a phrase that is used often, but is just as often misunderstood, or at least the full meaning is rarely appreciated. There is talk of "Integrated Product Teams (IPT)" and "Integrated Product and Process Development (IPPD)," but it is not always clear *what* is to be integrated. *American Heritage Dictionary* defines *integrate* as:

a) To make into a whole by bringing all parts together; unify

b) To join with something else; unite

Webster's New Collegiate Dictionary defines *integrate* as:

a) To form, coordinate, or blend into a functioning or unified whole; unite

b) To unite with something else

c) To incorporate into a larger unit

2.2.6.2 Integration of Project Elements

Basically, there are three types of things in a project that need to be integrated:

a) Organizational Elements,

b) End Product Elements, and

c) Enabling Product Elements.

Each project element must not only be integrated with the other elements of its own type, but must also be integrated with all the elements of the other two types as shown in Figure 2-7.

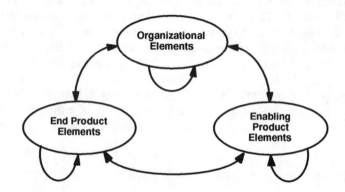

Figure 2-7 *Integration of Various Elements on a Project*

Organizational elements consist of such things as the people, facilities (buildings, telephone infrastructure, etc.), organizational structure, policies and procedures, computers, computer networks, services (mail, reproduction, legal, etc.), and so on. The end product and enabling product elements consist of various attributes of these products—requirements, functions, constraints—plus the products themselves.

2.2.6.3 Types of Integration

Table 2-4 shows a summary of the types of integration needed for a project and the activity that performs the integration. The items in the table dealt with by the SE Management activity are the *organizational elements*. The other items in the table are the *end product* and *enabling product elements* of a project.

As you can see, there are many things on a project that must be integrated. Most of these are the responsibility of the systems engineering process. Notice that nothing is said here about *who* performs the SE Process activities. They could be performed by project managers, engineering managers, project engineers, system engineers, or design engineers. The assignment of responsibility is the prerogative of the Developer (i.e., the development agency or organization). The important thing is that all relevant activities are actually assigned and performed.

Table 2-4   Mapping of Integration Items to SE Process Activities

What	Performing Activity
Disciplines	SE Management
Teams	SE Management
Projects (w/ themselves)	SE Management
Projects (w/ external entities)	SE Management
SE Tools	SE Management
Other Engineering Tools	SE Management
Methods (from different disciplines)	SE Management
Project Environment (w/ other environments)	SE Management
Products with Personnel (e.g., operators, installers, maintainers, etc.)	Synthesis
Requirements	Requirements Analysis
Functions	Functional Analysis
Products (Logical)	Synthesis
Products (Physical)	SI&V

2.3 Product Development

2.3.1 Development Project Environment

Figure 2-8 shows the development project environment for application of the SE Process. This environment establishes policy and procedures local to that project and develops project-specific plans. There may be project-specific tools that are applicable. Project management will have reviews of that project and will collect metrics to assess progress against plans and schedules.

Each development project resides within an enterprise which could be an organization within a company or an agency of the government. There will be multiple projects within that enterprise that compete for the resources of that enterprise. This environment also establishes policies and procedures. There may be standards and guidelines developed to aid each project in developing products. There are technologies (associated with the core competencies of that enterprise) that a project must draw upon. There is also the "company culture."

External to the enterprise are several factors that affect a development project: laws and regulations, legal liabilities, social responsibilities, technology base, labor pool, competing products, and industry standards.

All of these elements in the environments of a development project affect the manner in which a system is engineered. They must be recognized and understood for the influences they will have.

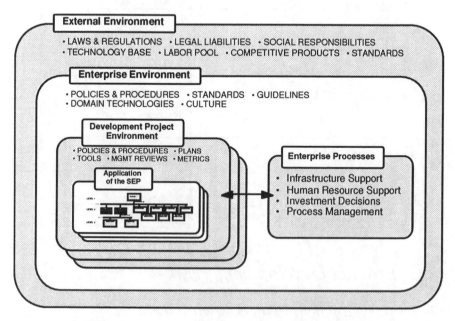

Figure 2-8   *Key Development Project Interfaces*

2.3.1.1 Enterprise Process Interfaces

There are enterprise processes that provide support to a project. These processes can have significant impact on the success of a project: infrastructure support, human resource support, investment decisions, and process management.

The infrastructure support process provides the following:
  a) shared database of corporate knowledge
  b) resource management (non-personnel resources such as facilities, land, vehicles, etc.)
  c) procurement (purchasing, stock room, etc.)
  d) research and development laboratories
  e) tools support (technical and non-technical)
  f) information management systems
  g) contracting
  h) security

The human resource support process provides the following:
  a) staffing
  b) employment services
  c) benefits
  d) training

The investment decisions process provides the following:
  a) new business development
  b) business case analysis
  c) strategic planning
  d) management reviews
  e) marketing

The process management process provides the following:
  a) development of process, methods, and tools
  b) acquisition of process, methods, and tools
  c) deployment and support of process, methods, and tools
  d) improvement of process, methods, and tools
  e) collection and analysis of process metrics
  f) recording and dissemination of lessons learned

2.3.1.2 Interface with Other Enterprise Projects

There will be other projects within the enterprise that will share resources, technologies, staff, etc. There may be several projects that are serving the same market segment. They may have similar product lines.

The development activities within a project need to be aware of any other project where there is potential conflict or potential gain from combining or coordinating efforts.

2.3.2 Acquirer/Supplier Relationships

Some of these other projects within an enterprise might be "subprojects" or "superprojects" for your project. Figure 2-9 shows the relationships between upper and lower layer projects. The higher level projects will flowdown requirements in a product specification to a subproject or, if the item is to be purchased, will flowdown the requirements in a procurement specification.

The systems engineering process will generate these specification documents and will need to manage the requirements. It is usually wise to involve the people from lower layer projects or from a component supplier in the requirements definition activity of the systems engineering process.

The "vehicle" of agreement could be anything from a formal contract to a handshake. Often these take the form of memoranda of agreement or understanding (MOA or MOU).

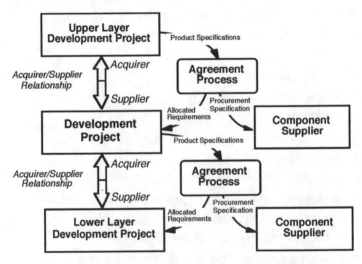

Figure 2-9 *Acquirer/Supplier Relationships*

2.3.3 Development Hierarchy

Figure 2-10 shows the relationship between the layers of development for a project. The lower layers could be developed within a single project or the work could be "contracted" out to subprojects, subcontractors, vendors, or suppliers. As the products at the lowest layers are built or acquired and tested, they are passed up to the next higher layer for integration.

This process continues until the entire system has been integrated and tested. Each step of integration must comply with the *development baseline* established for that level by the systems engineering process. Notice that the requirements for any layer not only come from an upper layer or the customer, but also from "local" stakeholders.

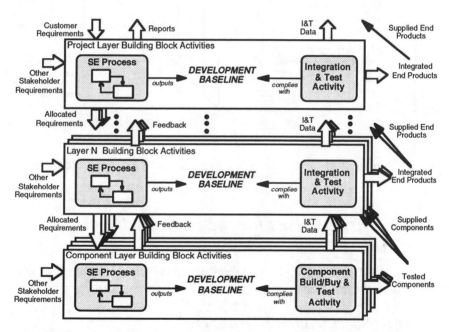

Figure 2-10   *SE Process in Context of Overall Development Activity*

2.3.4 Concurrent Development of Enabling Products

The enabling products for each building block may need to be developed or acquired if they do not already exist. Figure 2-11 shows how the enabling products must be integrated not only with the end products, but also with the other enabling products.

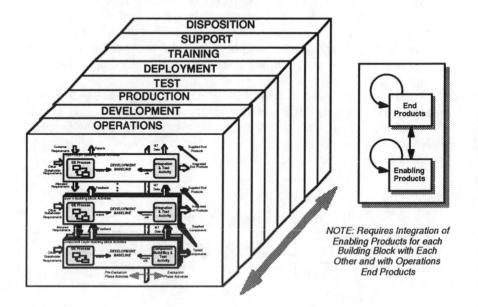

NOTE: Requires Integration of
Enabling Products for each
Building Block with Each
Other and with Operations
End Products

Figure 2-11   *Concurrent Development of Enabling Products*

2.4 Various Uses for the Building Block Structure

The building block defined above has many other uses on a project besides those already described.

2.4.1 Specification Hierarchy

The "specification tree" needs to be consistent with the building block as shown in Figure 2-12. The building block can be used to ensure that all the appropriate requirements for both end products and enabling products have been properly defined and documented.

The development of each enabling subsystem will take their requirements and transform them into specifications for the enabling products.

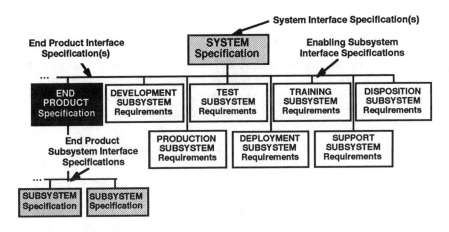

Figure 2-12 *Building Block Relationship to Product Specifications and Interface Specifications*

2.4.2 Technical Reviews

The various technical reviews associated with the building block elements are shown in Figure 2-13. The suggested order for having these reviews is from the bottom up.

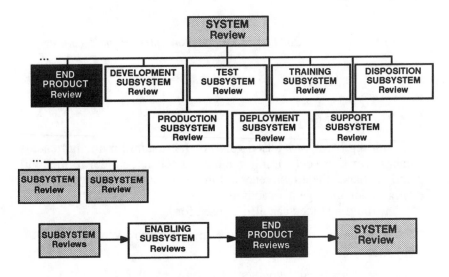

Figure 2-13 *Building Block Relationship to Technical Reviews*

2.4.3 Organizing Project Teams

Figure 2-14 shows how a project might be organized around the building block structure. Of course, a project will have other project teams than those shown See Chapter 11 for a description of other types of teams: leadership, cross-product, and cross-project teams.

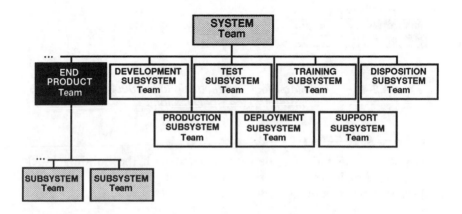

Figure 2-14 *Building Block Relationship to Project Teams*

2.4.4 Other Uses

The building block provides a very good structure for a requirements database for a project. Using a database schema similar to this will greatly enhance the effectiveness of the requirements management and configuration management activities.

The building block can also be used for:

a) defining the project tasks to be mapped into a Work Breakdown Structure (WBS)

b) assessing the development risk for a project

c) defining a process hierarchy

d) defining process metrics

2.5 The Nature of Requirements

2.5.1 Principles and Concepts

Some of the basic characteristics of requirements were defined in "What is a Requirement?", a paper written by the INCOSE Requirements Management Working Group [Harwell et al. 1993]. These requirements characteristics will not be repeated here, but some additional relevant concepts will be described. That paper is based on the concept that:

> *If it mandates that something must be accomplished, transformed, produced, provided [or constrained], it is a requirement—period.*[14]

This Section will not discuss how to write good requirements since this was covered in much detail in the paper, "Writing Good Requirements," a paper issued by the INCOSE Requirements Management Working Group [Hooks 1994]. However, some of the principles of requirements discussed herein will certainly assist the requirements analyst and writer in writing better requirements.

2.5.1.1 Requirements-Related Activities

It has been said that "Quality is conformance to requirements." [Crosby, 1979] Since quality is meeting the needs of the customer, then those needs must be requirements. Requirements development/management activities include the following:

a) elicit these requirements from customers and potential product/service users,

b) validate, definitize and prioritize these customer/user requirements,

c) specify these requirements in such a way that they can be implemented and verified,

d) identify alternative solutions to these requirements,

e) identify balanced and robust solutions that "best" meet these requirements,

f) verify that the implemented solution meets these requirements.

---

[14]The original paper omitted the word "constrained," but it was added later, according to Rich Harwell, one of the principal authors of that paper.

Usually the first two tasks above are not considered to be tasks relevant to systems engineering. This is so because traditionally these tasks have been performed by a contracting agency in the government (for government procured systems) or by the marketing organization (for commercial systems). However, if the systems engineer is to be successful now and in the future he needs to be more cognizant of the real needs of the user and customer. (The user here is meant to be the person who actually exercises the features of the system. The customer is the person with the money who purchases the system, often as an "advocate" for the user.)

2.5.1.2 The Problem Space Paradigm

Requirements essentially define the "problem space" whereas the design defines the "solution space." There may be more than one design that meets the requirements, but it is objective of systems engineering to find the most balanced and robust solution. (The word "optimal" is not used here since this implies a mathematical precision that is usually not achievable.)

Another way to think about requirements is to ask what questions the requirement must answer. The questions that may be addressed by any one requirement are:

a) *What* is needed? (Capability)

b) *What* is *not* allowed? (Constraint)

c) *How well* must this capability be performed? (Performance)

d) *Under what conditions* must this capability be performed?

Other ways to ask the "How Well" question include:

a) How fast or how often?

b) How many or how much?

c) How far or how long?

d) Et cetera.

Notice that the "How Well" criteria can either be a constraint (e.g., how much can the product cost?) or a capability (e.g., how much power to transmit?). The "Conditions" criteria can specify environmental conditions or such things as sequencing (capability A occurs immediately after capability B is completed) and timing (capability C must complete within 10 seconds).

2.5.2 Requirements Categories

Requirements can be divided into two basic categories:

a) Capabilities

    (1) a desired feature, trait, or faculty

    (2) an ability that leads to satisfaction of needs

b) Constraints

    (1) a constraining condition, agency, or force

    (2) a restriction or limitation outside one's control

Constraints keep us from providing capabilities unless they can be avoided or circumvented. These two requirement categories apply equally well whether the requirement is applied to a delivered <u>product or service</u> or to a <u>project task</u> (such as the task to design, build, or deploy a product or service.)

The balancing act in engineering a system is between:

a) Capabilities and Capabilities

b) Capabilities and Constraints

c) Constraints and Constraints

Figure 2-15 shows the relationship between capability and constraint type requirements. Capabilities that, when added to a system, "strain" against one or more of the system constraints are known as "design drivers." Sometimes a capability will be a "freebie" since it comes in under the umbrella of a more difficult capability. However, these so-called "freebies" are not always really free since they may violate one of the *unknown* constraints (such as ease-of-use). Often added features, even if "free," cost the user in terms of more difficult operation or increased training requirements.

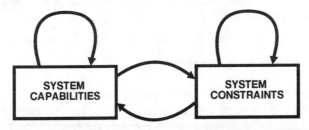

Figure 2-15   *Requirements Trade-Offs Between Capabilities and Constraints*

2.5.3 Requirements Life Cycle

One of the basic difficulties when dealing with requirements is to consider *all* requirements in the "whole universe" of the problem space. Figure 2-16 shows what might be considered the requirements life cycle, the whole universe of requirements, so to speak.

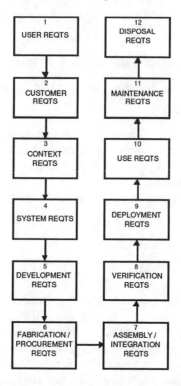

Figure 2-16   *The Requirements Life Cycle
(The Requirements Problem Space)*

There are two sets of life cycle requirements applicable to any development project:

a) product/service requirements set and

b) development requirements set.

The development requirements are the task requirements that apply to programmatic elements such as design, build, test, or deliver. These are often captured in contracts, statements of work, schedules and project plans. These have as much to do with project success as the requirements for the product or service under development.

The requirements life cycle may also represent the phases of a project, even though not all of these "phases" are normally considered as important to some projects. If we choose to ignore requirements from any of these areas, then we risk missing some of the critical requirements that might lead to system failure.

Failure in this context is not just the failure of a system component to perform its specified *function*, it could also be such things as failure to appeal to the customer's aesthetic sensibilities. (Of course, this could be thought of as not meeting one of a system's required aesthetic *functions*, but how often does systems engineering get involved in specifying of a system's aesthetic features? Maybe if they did, their systems would have greater success in the marketplace!)

Systems engineers often only concern themselves with the *system requirements* and sometimes even the *development requirements*. They leave the *user*, *customer* and *context requirements* to the marketing organization and mission analysts. They often leave most of the *development requirements* to be defined by design engineers and the *fabrication/procurement requirements* to the design and manufacturing engineers.

2.5.3.1 Role of Systems Engineering in the Requirements Life Cycle

Systems engineers are often overwhelmed by the sheer magnitude of just dealing with the *system requirements*, much less all the other requirements life cycle elements. Is it any wonder they would rather have other people handle the rest? But, if the systems engineer does not manage all these requirements, then who does? Often we find the really critical requirements are those that are lurking elsewhere in the problem space.

2.5.3.2 Requirements Flowdown

There are more requirements that must be dealt with than those that flowdown from the *system requirements*. Figure 2-17 shows the traceability that must be maintained between the system requirements and the requirements used for building the products and the system.

Figure 2-18 shows how the *users* of the end products will influence the system requirements. It is often best to keep the user requirements documented separately from the system requirements. There should be some latitude for changing system requirements as long as the impact on user requirements is known and understood.

User requirements are often documented by non-engineers. Therefore, they are often not in sufficiently technical form to be used directly by engineering. The systems engineering process should translate these non-technical requirements into specific language of the relevant engineering domain.

It is often as important to document who the users are as it is to document what they want. This may have great impact on how these user requirements are interpreted. You may need to define "use cases" or user scenarios. In other words, how will the users use the end products? What are their expectations? How will they respond to different situations?

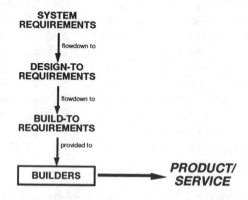

Figure 2-17   *Flowdown from System Requirements*

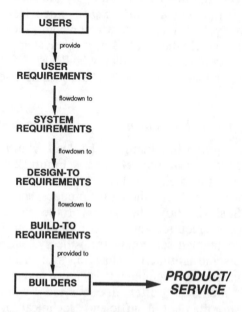

Figure 2-18   *Flowdown of Requirements from Users*

The users are of course very important since they will be the ultimate judge of success for your system. However, there are always other *stakeholders* (see Figure 2-19) for the system or product you are working on. They could be from manufacturing, test engineering, procurement, distribution, field support, etc.

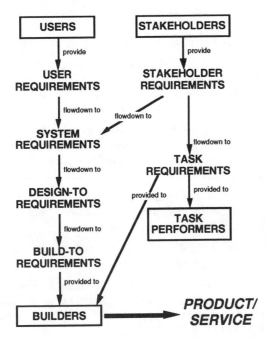

Figure 2-19   *Flowdown of Requirements from Stakeholders*

First, all stakeholders need to be identified. Then, the requirements from these stakeholders must be written down. These requirements must be prioritized and validated to ensure that they are indeed justifiable requirements. Many stakeholder requirements do not affect the product directly. There may be "task requirements" driven by stakeholders.

For example, your enterprise may have decided to standardize on a particular software language to save on development costs. They may also have defined a standard process to be used for software development. The software language standard directly affects the product, but it also may require some tasks to be performed in a certain manner. The software development process standard will definitely affect the development tasks—how they are performed, what tools to use, how to document decisions, etc.

Requirements are often not given due consideration. Perhaps this is because requirements are boring and design is fun. But, there is often more payback in getting the requirements right than getting the design right.

References

Crosby, P., *Quality is Free: The Art of Making Quality Certain.* McGraw Hill, 1979.

Harwell, Richard, et al., "What is a Requirement?" *Proceedings of the Third Annual International Symposium of the National Council on Systems Engineering (NCOSE)* , 1993.

Harwell, Richard, Jerry Lake, John Velman, and James Martin, "Anatomy of the Engineering of a System." *Proceedings of the Sixth Annual International Symposium of the National Council on Systems Engineering (NCOSE),* 1996.

Hooks, Ivy, "Writing Good Requirements." *Proceedings of the Fourth Annual International Symposium of the National Council on Systems Engineering (NCOSE)* , 1994, Volume 2.

IEEE, 1220 Standard, *IEEE Trial Use Standard for Application and Management of the Systems Engineering Process.* February 1995.

Martin, James N., "On the Relationship Between Enabling Subsystems and Enabling Products for the System." *Proceedings of the Sixth Annual International Symposium of the National Council on Systems Engineering (NCOSE),* 1996.

Martin, James N., "The PMTE Paradigm: Exploring the Relationship Between Systems Engineering Process and Tools." *Proceedings of the Fourth Annual International Symposium of the National Council on Systems Engineering (NCOSE),* 1994.

SESC Architecture Planning Group, *Toward a Recommended Practice for Architectural Description.* IEEE, 1996.

chapter three

Process concepts

"It requires a very unusual mind to undertake analysis of the obvious."
—Alfred North Whitehead

This Chapter explores the relationship between process and tools for systems engineering. There is much interest now in process maturity, yet there are many cases of processes being developed and deployed with little effect on the bottom line. In fact, there are times when the overall health of the engineering organization is not improved with the use of standardized processes and tools.

It is important to have a proper balance among process, methods, tools, and environment (PMTE) when performing systems engineering (SE) tasks. An improper balance leads to increased costs and lower quality, in addition to frustration for engineers and for management. This Chapter describes a paradigm for understanding the PMTE elements for systems engineering.

3.1 Relationship Between Process and Tools

One must understand the relationship between the SE process and SE tools in order to properly implement both (Figure 3-1). This Chapter explores this relationship.

| SYSTEMS ENGINEERING PROCESS | ⟶ | SYSTEMS ENGINEERING TOOLS |

Figure 3-1 *Jumping from Process to Tools*

SE *methods* bridge the gap between SE process and SE tools. With the increasing interest in continuous process improvement and assessment of process maturity for systems engineering, more attention needs to be given to the methods which support the process.

Quite often little consideration is given to using the proper *methods* when implementing a process and a set of tools on a project. The use of inappropriate methods can lead to inefficiencies and sometimes even failure.

3.1.1 Methods Compared

Systems engineering methods were compared and contrasted in the paper by (Armstrong 1993). He especially looked at some of the more recently popular methods such as object-oriented analysis (OOA) and quality function deployment (QFD). A major conclusion from that paper was that OOA and QFD "focus on significantly small subsets of the overall systems engineering analysis tasks. Consequently, serious shortfalls in overall understanding of the system may result if they are the only form of analysis used."

Armstrong's paper goes on further to say, "the better the systems engineer's understanding of a broader range of methods, the more effective their application will be and the better the systems engineering product."

3.1.2 Environmental Support for Tools and Methods

The right environment must be present to support the particular SE process, methods and tools used on a project. Some company environments are even hostile to the introduction of new methods and tools. These impediments to continuous process improvement must be overcome to achieve the higher quality products that good systems engineering practice will develop.

Most engineering textbooks only describe methods.[15] They usually do not describe the process, tools, or environment necessary for good engineering practice. The engineering student is exposed to some degree to process, tools, and environment in the classroom and laboratory, but not in a broad-based manner nor in sufficient depth. Therefore, the knowledge and skills of PMTE are often obtained on the job.

---

[15](Grady) and (Blanchard/Fabrycky) are exceptions to this since they cover SE process to some degree. (Blanchard 1991) and (Eisner) do a good job in covering SE tools. (Chestnut 1965) covers SE tools, but most of his "tools" are really methods according to the definition used in this book.

3.1.3 The Need for a Paradigm

There is generally not much guidance in the literature about the relationship among the PMTE elements. Even EIA/IS 632, an industry standard on systems engineering, only addresses process: "The scope and requirements of systems engineering are defined in terms of *what* should be accomplished, not *how* to accomplish it." [emphasis added] (EIA, 1994). To address this gap, this Chapter describes a PMTE paradigm,[16] a mental model used to screen data and filter new information with regard to SE process and tools.

3.2 The PMTE Paradigm

There is an intimate, supporting relationship among the PMTE elements shown in Figure 3-2. These elements must be consistent with each other, and must be well integrated and balanced to achieve the greatest benefit of good systems engineering practice. A process is executed using methods suitable for each process step. In other words, a particular process must be supported by certain methods. In turn, each method can be supported by one or more tools. A tool must be supported within a particular environment.

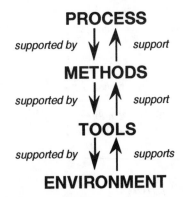

PROCESS

supported by ↓ ↑ *support*

METHODS

supported by ↓ ↑ *support*

TOOLS

supported by ↓ ↑ *supports*

ENVIRONMENT

Figure 3-2 *The PMTE Paradigm*

Figure 3-3 shows two other ways to visualize this concept. Without the proper environment as a foundation, any process cannot long stand. Process is at the core of the pie and at the pinnacle of the pyramid.

---

[16]The PMTE paradigm was first suggested to me by Jim Armstrong during a class on Advanced Systems Engineering. Mr. Armstrong first heard of the concept from Mr. Rudy Adler from the Software Technology Support Center at Hill AFB, Utah.

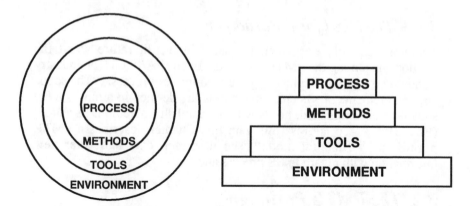

Figure 3-3   *The PMTE Pyramid and Pie*

3.3 PMTE Element Definitions

A few definitions are necessary to fully understand the PMTE paradigm. Briefly:

Process	*— Defines —*	What
Method	*— Defines —*	How
Tool	*— Enhances —*	What & How
Environment	*— Enables —* *(or Disables)*	What & How

3.3.1 Process Definition

A *process* is a logical sequence of tasks performed to achieve a particular objective. A process defines "WHAT" is to be done, without specifying "HOW" each task is to be performed.

The structure of a process provides several levels of aggregation to allow analysis and definition to be done at various levels of detail to support different decision-making needs. Figure 3-4 shows a generic process structure. A process phase consists of tasks, and a task consists of several steps. Other decompositions and levels of aggregation are possible.

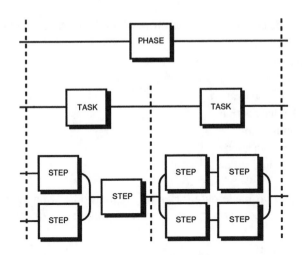

Figure 3-4 *A Generic Process Structure*

3.3.2 *Method Definition*

A *method* consists of techniques for performing a task, the "HOW" of each task.[17] (Even though a method is usually considered to be a process itself, for the purpose of definition in this Chapter, we will consider processes and methods to be separate and distinct.[18]) Methods usually imply a degree of discipline and orderliness. However, methods can be performed in an undisciplined fashion, even though good methods, in general, do enhance the structure and efficiency of a task.

All systems engineering methods deal with ideas. These ideas concern, among other things, functions, requirements, architecture, and verification. Methods have the following *attributes*:

    (a) Thought Patterns/Processes

    (b) Knowledge Base

    (c) Rules and Heuristics

    (d) Structure and Order

    (e) Notation

---

[17]"Method," "technique," "practice," and "procedure" in this context can be used interchangeably. A "methodology" can be defined as a collection of related methods, processes, and tools.

[18]At any level, process tasks are performed by using methods. However, each method is also a process itself, with a sequence of tasks to be performed for that particular method. In other words, the "HOW" at one level of abstraction becomes the "WHAT" at the next lower level.

All SE methods consist of one or more of the following *basic methods*:

    (a) Observation
    (b) Analysis
    (c) Synthesis
    (d) Conceptualization
    (e) Characterization
    (f) Optimization
    (g) Documentation
    (h) Communication

There are two basic categories of SE methods: Management and Engineering. Subcategories of each for Systems Engineering are as follows:

  a) Management
    (1) Planning
    (2) Organizing
    (3) Control
    (4) Direction
    (5) Integration
  b) Engineering
    (1) Requirements Analysis
    (2) Functional Analysis
       (or Structured Decomposition)
    (3) Architecture Synthesis
    (4) System Analysis and Optimization
    (5) System Element Integration and Verification
    (6) Engineering Documentation

3.3.3 Tool Definition

A *tool* is an instrument that, when applied to a particular method, can enhance the efficiency of a task. Of course, improper application of a tool will not likely increase efficiency. Most tools in the context of systems engineering are computer- and/or software-based. The purpose of a tool should be to facilitate the accomplishment of the "HOWs." A shovel is a tool which has little use to an operator who knows of no method for performing, for example, the process task of "Dig Hole."

SE tools can basically be categorized in the same manner as shown for methods above. However, many SE tools will fit into more than one category. For example, requirements management tools must span all methods categories since the purpose of requirements management is to establish and maintain all the proper relationships among the technical items (functions, requirements, architecture), along with the management items (project tasks, project requirements, personnel, project organization and facilities). Quite often, SE tools will work fairly well for either the technical side (the product) or the management side (the process), but not both. This shortcoming must be remedied for systems engineering to more effectively and efficiently bring complex products to the marketplace.

3.3.4 Environment Definition

An *environment* consists of the surroundings, the external objects, conditions, or factors, that influence the actions of an object, individual person or group. These conditions can be social, cultural, personal, physical, organizational, or functional. The purpose of a project environment should be to integrate and support the use of the tools and methods used on that project.

3.3.4.1 Environmental Categories

Environments can be categorized as follows:

a) **Computing**
   (1) Platform
   (2) Operating System
   (3) Application Software
   (4) Networks
   (5) Communication

b) **Communication**
   (1) Personal
   (2) Telephone
   (3) Video
   (4) Broadcast (Television, Telephone, etc.)
   (5) Mail
   (6) Electronic Mail (Email)
   (7) Internet

c) **Personal**
   (1) Teams
   (2) Work Groups
   (3) Personal Networks

d) **Organizational**
   (1) Project Organization
   (2) Functional Organization
   (3) Informal Power Structure
   (4) Agency Organization
   (5) Force Structure

e) Managerial
   (1) Policy and Procedures
   (2) Training
   (3) Funding
   (4) Commitment
f) Physical
   (1) Office
   (2) Home
   (3) On-the-Road (Hotel, Car, Plane, etc.)
g) System Life Cycle Environments
   (1) Development
   (2) Manufacturing
   (3) System Integration & Test
   (4) Deployment
   (5) Operations & Maintenance

3.3.4.2 The IPD Environment as a Special Case

Integrated Product Development (IPD) teams are relatively new elements in development environments today. The use of IPD teams requires new tools and methods to facilitate communication within each team and between teams. The use of IPD teams on a project was described in (Martin 1993).

So, one can see how the four elements of PMTE relate to each other and how they must be balanced and well integrated to ensure the efficient development of new, complex systems.

3.4 Distinction Between Process and Methods

3.4.1 Examples of Functional Analysis Process and Methods

The Functional Analysis process defined in this book consists of nine basic tasks (see Figure 3-5). The logical sequence of these tasks and several iteration loops are shown. This process diagram does not define "HOW" each task is performed. In fact, there are many methods for performing, for example, Task 403, "Define Functional Interfaces." Each of these "functional interface definition" methods has different ways of analyzing, synthesizing and documenting the interfaces. Each has its own notation and rules.

Many of the methods for performing functional analysis were described and compared in (Armstrong 1993). Some of the possible methods for performing functional interface definition are listed below:

a) IDEF Diagram
b) N x N (N2 or "N-Square") Diagram
c) Behavior Diagram
d) Functional Flow Block Diagram
e) Action Diagram
f) State/Mode Diagram
g) Process Flow Diagram
h) Function Hierarchy Diagram
i) Context Diagram

The method(s) that should be employed in a particular case depend on several factors:

a) Other methods used elsewhere in the process
b) Degree of integration with other methods used
c) Methods supported by existing or available tools
d) Experience level of engineers with a method
e) Ease of use
f) Ease of documentation of results
g) "Portability" between functional disciplines (e.g. systems, mechanical, electrical, software)
h) "Presentability" (How easily is the notation understood by the uninitiated? Is it easy to "see" the information or concepts when presented?)
i) How well documented[19]

Careful consideration should be given when methods are chosen for a particular project. And they should not be chosen just because they were used the last time! Even more important is the choice of SE tools. An improperly matched tool could actually decrease productivity and quality and increase costs.

---

[19]Some methods are not well documented. These might be newly developed, or handed down verbally to new engineers, or "kept in the closet" for reasons of job security. Some methods may be documented, but "no one does it that way anymore." Often the engineering procedures manual for a company is not kept current with the latest methods.

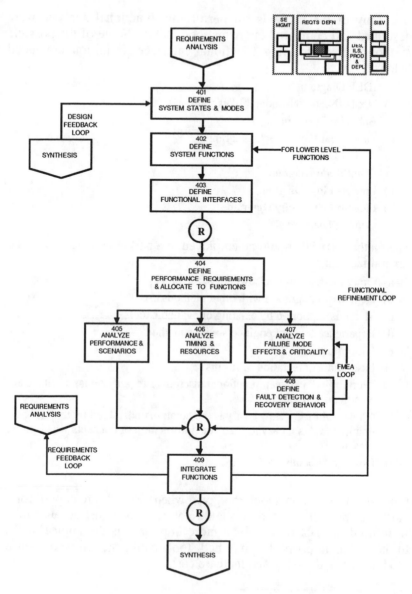

Figure 3-5 *A Functional Analysis/Allocation Process*

3.4.2 Object Oriented Analysis and Design (OOA/OOD) Methods

Object Oriented Analysis and Design has been touted as the "new process" of doing systems engineering. In reality, OOA and OOD are merely a collection of new (and some old) methods[20] that support the SE process. Like any other method, these need to be applied in the proper manner. OOA/OOD is difficult to apply without tools that support the particular rules and notation of a particular OO technique. There are many different OO methods and tools available. These must be analyzed carefully to ensure their proper application. OO methods do not, and cannot (except maybe in some special cases), replace standard functional analysis techniques, but rather supplement them. Even the objects themselves must eventually be analyzed for their proper functional behavior.

3.4.3 Role of Heuristics

Systems architecting has gained a following lately and is being espoused by some as a new paradigm for doing systems engineering. Use of heuristics as described in Eberhardt Rechtin's books, *Systems Architecting* (Rechtin 1991) and *The Art of Systems Architecting* (Rechtin 1996), is a method for performing systems engineering tasks. His book contains many heuristics that can be applied throughout the SE process. Heuristics are especially useful when there is a lack of information which might normally be required when using a normal technique or method.

Heuristics are usually used at a higher level of abstraction during the SE process than non-heuristic techniques. So far there are few, if any, tools that support heuristics. One possibility might be to put Rechtin's book into a hypertext-like database which would pull up relevant heuristics similar to the way electronic word thesauruses can give you "synonymous" words when given a "seed" thought.

---

[20]*Roget's Thesaurus* is an example of a very old use of object-oriented classification and organizational techniques.

3.5 Systems Engineering Development Environment (SEDE)

SE methods and tools must be firmly supported by the proper environment and process to have effective and efficient systems engineering for a project. It is usually the role of SE to ensure that all processes and environments on a project are well integrated and used properly. This role needs to be expanded to include methods and tools. This is the purpose of a Systems Engineering Development Environment (SEDE) Plan.

3.5.1 Management Support of the SEDE

As shown in Figure 3-6, methods and tools must be firmly supported by the proper environment and process. Management's role is to ensure that engineering follows the proper process and that the environment is supportive.

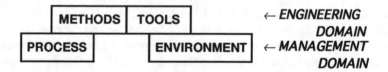

Figure 3-6 *Process and Environment Form the Basis for Good Execution of Systems Engineering*

Management should also give guidance to engineering on the proper (engineering) methods and tools and ensure that the engineers are adequately trained in them. Also, a *manager's* management should give guidance to the manager on the proper (management) methods and tools and ensure that the managers are adequately trained in them. (In this sense, the first-line managers should be acting as the "engineers" for the process. In other words, they should be "systems engineering" the project.)

3.5.2 Adjustment of the PMTE Model for an Organization

When any of the PMTE elements in a company or on a project are going to be changed, one must consider the transition path. Some paths are more difficult due to the organizational and personal inertia.

As shown in Figure 3-7, the PMTE elements lower in the "pyramid" are more stable, thus harder to change. In fact, often one encounters a "religious zeal" from the people who are the current stakeholders for existing methods, tools and environments. This is the management challenge: Changing any of the PMTE elements requires the wisdom of Solomon and the fortitude of Hercules.

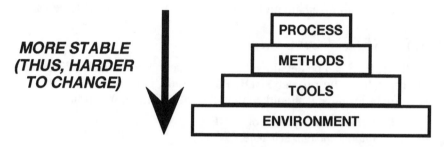

Figure 3-7   *PMTE Pyramid Stability Often Leads to Stagnation*

3.5.3 Relationship Between the SE Management Plan and the SEDE Plan

As shown in Figure 3-8, most SE Management Plans do not entirely cover the full spectrum of the PMTE model. Some SEMPs cover SE tools, but they often do not define the methods to be used with each tool nor do they even define the "manual" methods to be used. The SEDE Plan should cover the gaps that exist in the project SEMP (or possibly even in the corporate "generic" SEMP).

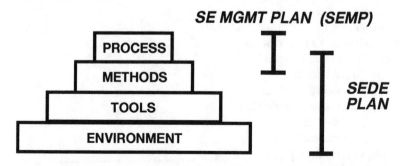

Figure 3-8 *The SEDE Plan Needs to Fill the Gap*

3.5.4 Tailoring the SE Development Environment

The SEDE consists of all the PMTE elements required to execute a particular job. A generic SEDE will often not be sufficient for all projects. Some tailoring of the SEDE is usually required and sometimes new PMTE elements must be developed specifically for a project. Off-the-shelf PMTE elements are, of course, desirable whenever they truly meet the needs of the project. SEDE elements that are easily customized are preferable to static, unchangeable elements.

3.5.5 SEDE Plan Contents

A SEDE Plan can be developed for a company and for each project, as appropriate. The project SEDE Plan should document the project-specific tailoring and should contain the following information:

1. **Control and Integration of SEDE**
   a) Implementation Schedule
   b) Engineering Policy
   c) Corporate Agreements
   d) Management Structure
   e) Team Structure
   f) Training Plan
   g) Change Control
2. **Process Description**
   a) Tailoring of SE Process
   b) Interface with other Processes
3. **Methods Description**
   a) Process/Method Correlation Matrix
4. **Tools Description**
   a) Computer Platforms
   b) Computer Software
   c) Computer Networks
   d) Other Tools
   e) Process/Tool Correlation Matrix
5. **Environment Description**
   a) Computer Facilities
   b) Office Facilities
   c) Laboratory Facilities
   d) Communication Facilities
      (1) Telephone
      (2) Videoconference
      (3) Networking
      (4) Electronic Messaging and File Transfer
      (5) Physical Messaging (Non-Electronic)

3.6 Mapping PMTE Relationships

3.6.1 PMTE Gap Analysis

If the supporting links among *process* tasks, *methods, tools* and *environments* for a particular project or organization are analyzed (see Figure 3-9), one may discover several interesting gaps in a set of PMTE elements under study. There may be "orphaned" methods that have no tools supporting them. This may not necessarily be bad, but nevertheless, these links need to be understood if a balanced and integrated set of PMTE elements is to be used on a project or within a company.

Some methods are simple enough that a sophisticated tool may not be needed. For example, brainstorming is a method that is useful for almost all process tasks, but it is usually simple enough to be implemented using very simple tools such as pencil, paper, blackboard, etc. However, some more sophisticated versions of brainstorming, such as "mind mapping," could benefit from a more automated approach. In fact, some tools on the market now specifically support mind mapping and other complicated brainstorming methods.

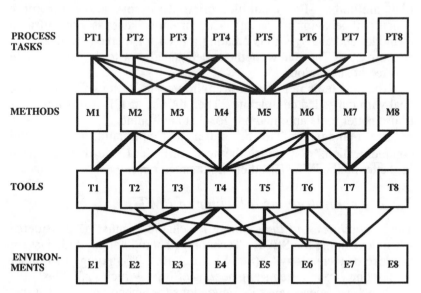

The thickness of the lines indicates the degree of support or compatibility between elements.

Figure 3-9 *Mapping Among the PMTE Elements to Perform Gap Analysis*

3.6.2 Process Improvements

By using the PMTE mapping approach, one can determine where new methods are applicable in the processes a company uses and make sure that these methods are used in appropriate situations. One might also discover that some of the tools support methods that are not understood by the practicing systems engineers. If they are not understood, they will not be used. Thus, opportunities for training in certain methods might be identified.

One might also discover some incompatibilities between tools and environments. For example, many of the more sophisticated tools run on workstations and many companies have yet to deploy workstations widely. Some of the tools support client/server modes of operation, yet this may require investment in additional computer networks to efficiently use this feature.

3.7 Roles of Technology and People

3.7.1 The Technology Ripple Effect

New technology in the products being developed forces us to develop new methods. For example, microprocessors allow for more concurrent behavior in systems. Concurrent behavior is more difficult to model and analyze, thus creating a need for methods to deal with concurrency. These new methods demanded new tools, especially in the area of simulation.

The new tools required new environments, such as IPD teams, to deal with the added complexity. Then the use of IPD teams forced us to use better tools and methods for that environment. IPD teams also require management to be better at "command, control and communication." Of course the need for better C^3 leads to the need for better methods, tools and environment, *ad infinitum*.

3.7.2 Technology as a Driver of PMTE

The *capabilities and limitations* of technology must be considered when developing the SEDE. Technology should not be used "just for the sake of technology." (See Figure 3-10.) Technology can either help or hinder our systems engineering efforts. Which shall it be?

Math and science are the foundations for technology. Notice that technology can be applied to the PMTE elements to enhance their effectiveness. For example, computers can be added to the environment to enhance productivity and communication. Object-oriented theories can be applied to tools such as software compilers and software CASE tools. Fractal theory can be applied to methods to enhance the capabilities of those methods. Hypertext concepts can be applied to the process to assist in the definition and use of those processes.

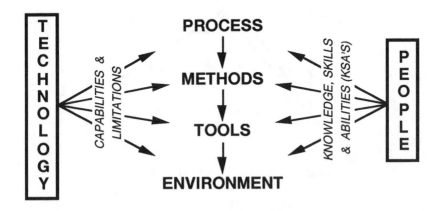

Figure 3-10 *Effects of Technology and People on the PMTE Elements*

3.7.3 People are the Cornerstone of Systems Engineering Practice

When choosing the right mix of PMTE elements, one must consider the *knowledge, skills and abilities* (KSA) of the people involved. When new PMTE elements are used, often the KSAs of the people must be enhanced through special training or special assignments.

Figure 3-11 shows the effects of time on KSAs for an individual. Ability is more or less constant throughout one's life, while knowledge generally increases. Skills, such as using certain tools, usually peak soon after college and decrease from then on, unless special care is taken to keep these skills current. Buying expensive tools without providing training to use them most often does not improve the development process, since the tools may be misused, or not used at all.

Some corporate cultures assume that knowledge and skills are not necessarily relevant as long as they hire the smartest engineers around. Unfortunately, high grade point average does not necessarily correspond to excellent SE skills and knowledge.

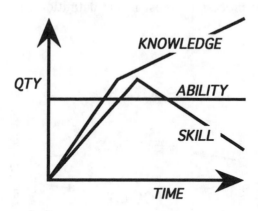

Figure 3-11   *Effects of Time on KSAs*

3.7.4 Creativity vs. Structure

Traditionally, most systems engineers have been trained as "design" engineers and pride themselves in being very creative. Often creativity is thought to be antithetical to the "structured thinking" skills and abilities so necessary for good SE practice. In reality, SE needs to bring "structure" to a problem so that creativity can be more properly focused. Often the systems engineer must be extremely creative in "discovering" the structure for a problem and in communicating this structured set of information to both the potential customer and the design team.

Many engineering students learn little about "structuring" a problem since most professors do this for them. The student is merely challenged to "solve the problem" that the professor has already "defined" for them. Is it any wonder that few "structure-minded" systems engineers come out of our universities? Real world problems do not often come as well structured as those in the classroom.

3.8 Implications for Management

The concepts described in this Chapter have been used successfully within many organizations to help management understand the proper role of tools in the entire environment of process and methods. For example, a process improvement survey was conducted using personal interviews of systems engineers and their managers. The survey responses were mapped into the PMTE categories and this greatly enhanced the usefulness of the survey results. It was found to be easier to identify specific process improvements and target these for action plans.

The methods to be used by systems engineers in a project and company should be well documented. These methods often need to be tailored to the specific needs of a project. If everyone on a project uses the chosen set of methods in a consistent manner, this leads to greater communication among the project team and with the customer, fewer technical errors due to miscommunication, and greater productivity and quality.

The process, methods, tools, and environment used on a project and within a company need to be established through engineering policy. The particular methods should be documented in engineering procedures documents. Whenever possible, reference can be made to a particular method as documented in reference handbooks and textbooks.

The need for proper training in the use of tools and methods is often overlooked by management. How often has a tool been blamed for a project failure, when in reality the people involved did not use the tool in the proper manner? Sometimes engineers use the methods that their tool supports even though these may be the wrong methods for that phase of the project, or that type of technology, etc. Even worse is using a tool to perform a method that is not supported by that tool.[21] Training must be considered as part of the investment equation when implementing tools and methods in a project.

Introducing a new process, method, tool or environment to a project presents significant risk and this risk must be mitigated to an acceptable level. Risk can be reduced by implementing new elements on a small pilot project. However, caution is advised since PMTE elements sometimes do not scale up linearly.

Systems engineering should lead the planning of the PMTE elements to be used on a project. Systems engineering should also ensure that these elements are integrated with each other in a balanced and robust manner and that they are integrated with the non-engineering elements such as business, project management, manufacturing, deployment, etc.

It is important to have a *proper balance* among process, methods, tools, and environment when performing systems engineering tasks. An improper balance leads to increased costs and lower quality, in addition to frustration for engineers and for management.

---

[21] As was said at a recent INCOSE meeting, "A fool with a tool is still a fool."

References

Armstrong, James R., "Systems Engineering Methods Compared." *Proceedings of the Third Annual International Symposium of the National Council on Systems Engineering (NCOSE)* (July 26-28, 1993).

Blanchard, B., and W. Fabrycky, *Systems Engineering and Analysis.* Prentice Hall, 1990.

Blanchard, B., *System Engineering Management.* John Wiley, 1991.

Chestnut, Harold, *Systems Engineering Tools.* John Wiley, 1965.

EIA/IS 632, *Systems Engineering.* EIA, 1994.

Eisner, H., *Computer-Aided Systems Engineering.* Prentice Hall, 1988.

Grady, Jeffrey O., *System Requirements Analysis.* McGraw Hill, 1993.

Martin, James N., "Managing Integrated Product Teams During the Requirements Definition Phase." *Proceedings of the Third Annual International Symposium of the National Council on Systems Engineering (NCOSE)* (July 26-28, 1993).

Rechtin, Eberhardt, *Systems Architecting: Creating and Building Complex Systems.* Prentice Hall, 1991.

chapter four

Systems engineering process overview

"There is a certain method in his madness."

—Horace

4.1 SE Process Roles

Four types of teams are typically used to perform the SE Process activities defined in this book:

1) SE Management Team

2) Requirements and Architecture Team

3) Development Team

4) System Integration and Verification Team

The Development Teams[22] receive technical management direction from the SE Management Team[23] and these teams report technical

---

[22]The development teams consist of designers (software, electrical, mechanical, materials, etc.), ILS engineers, and manufacturing and deployment engineers. Systems engineers should be integral members of the high-level development teams. There may be representation from other disciplines including specialty engineering. Section 4.3.4 discusses the tasks of these teams.

[23]The SE Management team will typically have representatives from systems engineering, design, ILS, manufacturing and deployment, as appropriate. It may also have representatives from an external customer or from marketing.

progress back to the management team. The Development Teams also receive their development requirements from the Requirements/ Architecture Team and they then provide their product and process designs and prototypes to the System Integration and Verification Team for design qualification and user validation.

Chapter 11 describes how the project organization and teaming structure should be established consistent with the system architecture. Also described are cross-product and cross-project teams for addressing such issues that are outside the scope of a single product team.

4.2 Systems Engineering Process

Figure 4-1 depicts a top level view of the SE process. The process comprises the following three subprocesses with interfaces to the design, ILS, production and deployment processes:

- SE Management Subprocess

- Requirements and Architecture Definition Subprocess

- System Integration & Verification (SI&V) Subprocess.

These subprocesses are described further in Section 4.3. Interfaces and key process relationships are included herein, as necessary, for clarity and understanding of the SE process. A detailed discussion of the SE process and process steps is included in Chapters 5 through 8.

The SE process is defined for a complete development (lust-to-dust) effort, starting from general requirements and ending with a compliant product or process. A complete project begins with a generally stated customer need and is completed when the system is delivered to the customer and successfully integrated into the customer environment.

The original source of many system requirements for development projects is the User Requirements Document (URD). The URD is a document that specifies the minimum acceptable performance requirements of a system stated in terms of operational and performance capabilities. The URD may be prepared by a marketing organization, a user agency, or by the users themselves. The user requirements are often not stated in technical terms. The tasks of the SE Process will transform these user requirements into technical requirements that are verifiable and are stated in the specific language of the affected engineering domain.

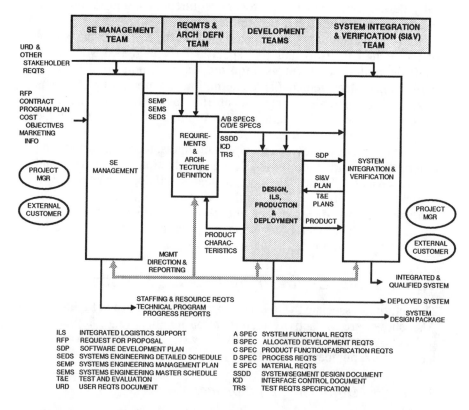

Figure 4-1 *Overview of the Systems Engineering (SE) Process*

The systems engineering process is executed by numerous participants, both external and internal to the engineering organization. Participants within this organization include traditional hardware and software design engineers, systems engineers, test engineers, and system analysts. External participants consist of project management, manufacturing, deployment, logistics, field support, specialty engineers and external customers. This multi-disciplinary approach is inherent in the systems engineering process and is consistent with the concept of teaming and integrated product development addressed in Appendix B.

The SE effort on a project or contract may begin at any point in the process, depending on what has been accomplished in previous phases, contracts, or proposal preparation activities.

The various disciplines, such as design, manufacturing, test, deployment, specialty engineering, and logistics, use the SE process to fully define the product and process requirements, and to agree on interfaces needed to meet those requirements.

Risk Management is a key component in the overall management of the technical effort and is a specific item in the SEMP. Risk analysis and assessment is required as an input to Technical Performance Measurement (TPM) definition and monitoring. Systems engineering provides test traceability back to the customer requirements, while verification in SI&V closes the loop between the design and the requirements.

The SE process provides requirement inputs, technical monitoring criteria and coordination for the design, production, deployment, and ILS processes. These development teams will perform their own requirements analysis specific to their discipline and coordinate their analysis, design, development and test with SI&V and specialty engineering. The overall development process will be managed by those who perform the SE Management tasks.

4.3 Systems Engineering Process Elements

The activities associated with each subprocess are shown in Figure 4-2. Also shown are the activity identification numbers (e.g., 500). The tasks associated with each activity will have their own task identification numbers (e.g., 503).

4.3.1 SE Management Subprocess

This subprocess contains the activities associated with management of the technical effort on a project. This includes the development activities of design, ILS, production and deployment. It involves planning the technical management of the project, controlling the technical activities, and integrating the efforts of all involved in the execution. These activities provide the foundation for the integrated product development teams. The SEMP is the basis for the management and execution of the project (see Section 4.4).

Essential elements of this subprocess are as follows:

- Overall technical program objectives will be defined.

- Technical management processes are selected and tailored, including methods and techniques for effectiveness and risk analysis, technology insertion, TPM, and ILS.

- The SE process as applied to the project will be documented in a SEMP/SEMS which will be coordinated, distributed, and used for controlling and integrating the technical program.

- Project technical progress will be tracked and managed.

- Configuration Management (CM) will control configuration changes and maintain applicable technical documentation.

- Risk Management will be an ongoing activity.

- Product development teams have proper and timely representation from all relevant disciplines, including from specialty engineering.

4.3.2 Requirements and Architecture Definition Subprocess

This subprocess contains the activities associated with transforming the customer requirements into documented technical requirements and an architecture at the system, segment, subsystem, system element, item, component, or unit levels.[24] This is an iterative process, and the steps will be essentially the same at each level of design detail. The purpose is to provide a complete set of traceable requirements for the products and processes at all levels to ensure that design is focused on the customer's needs.

Essential elements of this subprocess are as follows:

- Customer needs and requirements are defined or refined at the start of this activity.

- Technology is assessed during synthesis to determine the constraints on the entire Requirements and Architecture Definition Subprocess.

- System operational and support concepts are defined/refined.

- Measures of effectiveness are defined in terms of the mission profile and customer needs.

- Requirements are analyzed, and are derived and further refined where necessary.

- Optimization analysis identifies desired characteristics.

- System behavior is defined through functional analysis, and functional performance requirements are allocated to these functions.

---

[24]These names given to the levels in the system hierarchy are for illustration purposes only. Each project, company, organization, and industry will have its own unique designators for the system hierarchy levels.

- Architecture is defined through synthesis and requirements are traced to all system elements.

- Trade studies analyze alternatives and support selection of a balanced, optimized set of system elements.

- Requirements for each system element are documented in specifications, drawings and interface documents and placed under Configuration Management control.

- The results of this subprocess are presented and agreed upon at the appropriate major technical review.

4.3.3 System Integration and Verification (SI&V) Subprocess

This subprocess contains the activities associated with integration and verification of all subsystems,[25] which results in a completely integrated and qualified system. These include establishing the integration plan and schedule and the associated integration test cases; developing integration and test procedures; and performing system integration, integration testing, design qualification testing, product acceptance testing, and process qualification testing.

Essential elements of this subprocess are as follows:

- A System Integration and Verification (SI&V) Plan is developed which defines the overall test program and the verification method and level for all technical requirements. The SI&V Plan includes the test philosophy and guidelines. Special test equipment, facilities and software are also defined in the SI&V Plan.

- Test and evaluation requirements are defined and included in Section 4 of the appropriate system or element level specification. A Requirements Verification Matrix (RVM) is developed, and verification procedures are defined, where applicable.

- Testability of the requirements is established at the requirements and design reviews.

- Integration and test activities are developed and documented in test plans for each system element. Factory Acceptance Test (FAT) and Site Acceptance Test (SAT) plans and requirements are developed.

---

[25]The subsystems to be integrated could be at any level in the system hierarchy. The actual level in the hierarchy where SI&V is responsible should be defined in the SEMP and SEDS.

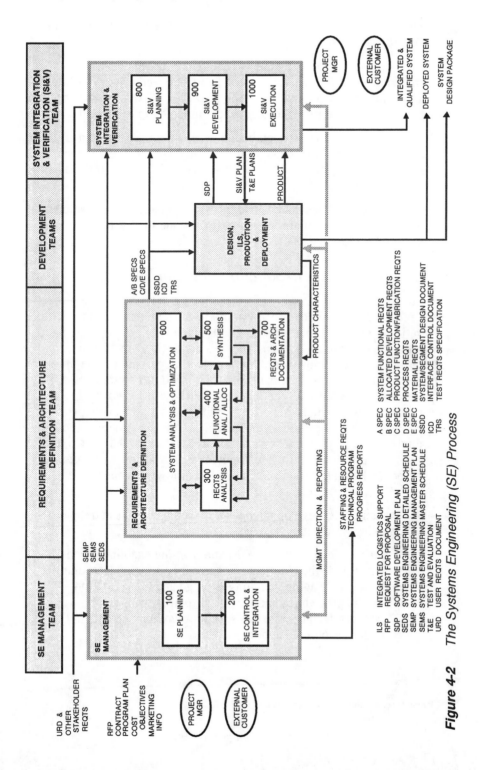

Figure 4-2 *The Systems Engineering (SE) Process*

- The adequacy of the design and readiness for test are reviewed at a Test Readiness Review (TRR).

- Tests are performed using the Test, Analyze and Fix (TAAF) approach. Test failures are recorded and analyzed using a Failure Reporting and Corrective Action System (FRACAS). Retests are performed as required. Results are documented in test reports.

- The adequacy of the production and deployment processes, and their readiness for full-scale production/deployment are reviewed at a Production Readiness Review (PRR) and a Deployment Readiness Review (DRR).

- Completion of all test and design qualification activity is ensured at the Functional Configuration Audit (FCA) and the System Verification Review (SVR).

4.3.4 Design, ILS, Production and Deployment

There are usually *eight separate systems* being developed for a project that must be integrated into a complete system that meets all the operational requirements. These eight systems are those associated with the *eight primary system functions* as defined in EIA 632 and IEEE 1220: development, production, test, deployment, operations, training, support, and disposition. The products associated with operations are called *end products* (since they are used by the end user), and the products associated with the other seven primary functions are called *enabling products*[26] (since they "enable" the end products to be put into service, kept in service, and taken out of service). The four types of development teams shown in Figure 4-2 are associated with the eight systems in the following manner:

1) Design teams develop the *operations* end products and the *test* enabling products.

2) ILS teams develop the *support* and *training* enabling products, which include the maintenance and support facilities, training materials, packaging, handling, storage and transportation (PHS&T) procedures and facilities, sparing plans and procedures, technical manuals, support equipment, and computer resources support.

---

[26]Examples of enabling products are given in Section 2.2.5.1.

3) Production teams develop the *production* enabling products, which include the production facilities, manufacturing work instructions, production test facilities and equipment, and procurement procedures.

4) Deployment teams develop the *deployment* and *disposition* enabling products, which include the deployment procedures and facilities, transition plan, special installation tools, disposal facilities, and installation and checkout procedures and tools. These also include the administrative procedures and equipment for keeping track of the products and their characteristics throughout their life.

5) Management teams and other teams develop the *development* enabling products, such as engineering plans and schedules, development prototypes, analysis tools, and laboratory facilities and equipment.

These functional areas receive specifications, requirements, interface definition and other design documents from the SE process. They participate in the SE Management, Requirements and Architecture Definition and SI&V activities as members of Integrated Product Teams (IPTs). For example, all are involved in the iterative development and review of system requirements, and provide coordination and verification of requirements as members of the Requirements and Architecture Definition team.

In planning their efforts, these functional areas help produce key project plans such as the Software Development Plan (SDP), ILS plan, producibility plan, deployability plan, etc. In the Requirements and Architecture Definition Subprocess, they are involved in the team effort to develop specifications and documentation which are inputs into their own functional processes. These functional development teams develop a design that meets these specifications and deliver a product to the SI&V process for design qualification testing and integration of all system elements into a complete system.

Each functional area listed will have its own process. The involvement shown here is the interface with the SE process. Specific interfaces will be further described in Chapters 5 through 8.

Essential elements of this relationship are as follows:

• Participating in the development of plans

• Assisting in system-level requirements analysis and definition

• Participating in the IPT development of system, segment, and other level specifications

- Participating in the iterative optimization loop of all the above listed elements as part of the integrated development process

- Producing plans, documentation and products to deliver to System Integration and Verification Subprocess

- Producing plans, documentation and processes for manufacturing, verification, deployment and support of the system products

4.4 Systems Engineering Management Plan (SEMP)

The Systems Engineering Management Plan (SEMP) is the primary, top level technical management document for the integration of all engineering activities within the context of, and as an expansion of, the project plan. A SEMP should be prepared for each project and regularly updated as development progresses.

The SEMP is not necessarily a long document. For some projects, it could be a page long, for others it could be hundreds of pages long. The plan needs to be specific to the needs of that project. It needs to be a "living" document, updated as often as needed as new information becomes available. It is often best if the SEMP references existing organizational policies and procedures. There is no need to duplicate what already exists.

4.4.1 SEMP Contents

The contents of the SEMP are described in EIA 632[27] and may include three parts as depicted in Figure 4-3 and described as follows:

Part I, Technical Program Planning and Control, describes the proposed process for planning and control of the engineering efforts for the system's design, development, test and evaluation.

Part II, Systems Engineering Process, includes specific tailoring of the SE process as described in this book, implementation procedures,

---

[27]The SEMP described in EIA 632 no longer has 3 parts, but the contents are essentially the same with more emphasis placed on multi-discipline teaming and technology transition.

trade study methodology, types of models to be used for system and cost effectiveness evaluations, generation of applicable documentation and specifications.

Part III, Engineering Specialty Integration, describes the integration of technical discipline efforts and parameters into the SE process and includes a summary of each technical discipline effort with a cross reference to the specific plan.

The SEMP forms the foundation for all engineering activities during the entire project and is the means for documenting the tailored SE approach to be used for a specific project. The development of the SEMP is a systems engineering management responsibility, but it must reflect the combined, coordinated inputs of the Project Manager and all other participants in the project.

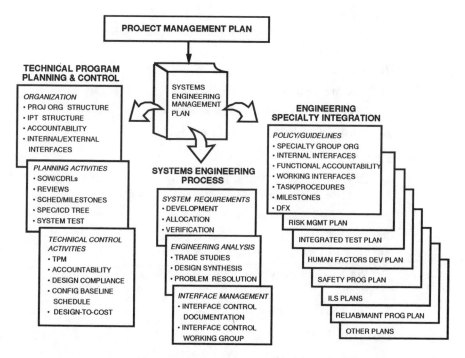

Figure 4-3   Systems Engineering Management Plan (SEMP)

4.4.2 SEMP Checklist

A comprehensive and well-thought-out SEMP is the key element in the planning of the systems engineering process. The SEMP should address the following questions:

a) Problem

1) What is the problem we're trying to solve?

2) What are the influencing factors?

3) What are the critical questions?

4) What are the overall project constraints in terms of cost, schedule, and technical performance?

5) How will we know when we have adequately defined the problem?

6) How will we know when we have adequately solved the problem?

7) Who are the customers?

8) Who are the users?

9) What are the customer and user priorities?

10) What is the relationship to other projects?

b) People

1) How are we going to structure the project to enable this problem to be solved on schedule and within cost?

2) What does systems engineering management bring to the table?

3) How will we integrate the various disciplines?

4) What special knowledge, skills, and abilities will be required?

5) What training is needed?

c) Information

1) What metrics will be used to measure technical progress?

2) What metrics will be used to identify process improvement opportunities?

3) How will we measure progress against the plans and schedules?

4) How often will progress be reported? reported by whom? to whom?

5) How will we assess risk? what thresholds do we need for triggering mitigation activities? how will we integrate risk management into the technical decision process?

6) How will we communicate across and outside the project?

7) How will we record decisions? where?

8) How can we incorporate lessons learned from other projects?

d) Process

1) What is our systems engineering process for this project?

2) What are the methods that we will apply for each systems engineering task?

3) What are the tools we will use to support these methods? how will the tools be integrated?

4) How will we control configuration development?

5) How/when will we conduct technical reviews?

6) How will we establish the need for and manage trade-off studies?

7) Who has authorization for technical change control?

8) How will we manage requirements? interfaces? documentation?

e) Technology

1) How and when will we insert new or special technology into the project?

2) What is our relationship to research and development labs? how will they support us? how will we incorporate their results?

3) How will we incorporate system elements provided by the customer or user? how will we certify the adequacy of these items?

4) What facilities are required?

5) When/how will we transition to Production?

6) When/how will we transition to Product Support?

4.5 *Systems Engineering Master Schedule (SEMS) and Systems Engineering Detailed Schedule (SEDS)*

The Systems Engineering Master Schedule (SEMS) documents the accomplishment criteria for all critical tasks and milestones in the SEMP and the Systems Engineering Detailed Schedule (SEDS).[28] The SEMS uses metrics that include Technical Parameters (TPs)[29] and specific success criteria for progress assessment. The interrelationships between the SEMS, SEDS, WBS, SOW and requirements are shown in Figure 4-4.

Since the SEMS does not usually have programmatic events (such as project management reviews), there is usually a higher level master schedule that contains these items. The SEMS needs to be kept consistent with the higher level project master schedule.

The SEMS is usually submitted with a proposal. The SEMS (and high-level SEDS) may also be contractual, implemented through the Statement of Work (SOW), and included in the reporting items in the Contract Data Requirements List (CDRL). When this is the case, a contract modification is required for all changes to the SEMS.

Key purposes and characteristics of the Systems Engineering Master Schedule are to:

- Provide for contractual/ payment incentives to meet the quantitative criteria identified in the SEMS

- Serve as a process control tool to ensure task accomplishment

- Provide project management with progress criteria

- Reduce development risk

- Instill design discipline

---

[28]The SEDS defines "when" the tasks are to be done while the SEMS defines "what" needs to be done in terms of completion criteria for each major task and milestone. The SEDS is equivalent to what many projects call a "project master schedule", except that the SEDS usually does not have programmatic events (such as project reviews).

[29]TPs are a selected subset of the system's technical metrics tracked in Technical Performance Measurement (TPM) monitoring activity. Critical TPs are identified from risk analyses and contract specification or incentivization, and are designated by management.

- Support decision milestones by —

  - Requiring up-front agreement with the customer on key tasks, associated schedules, and success criteria

  - Providing progress based criteria (a) by being event-oriented (not calendar-based) and (b) for strengthening the role of technical reviews and audits

The SEMS is supported by the Systems Engineering Detailed Schedule (SEDS). The SEDS identifies the detailed technical tasks required to accomplish each task or event in the SEMS. The SEDS is *time-based* while the SEMS is *event-based*.

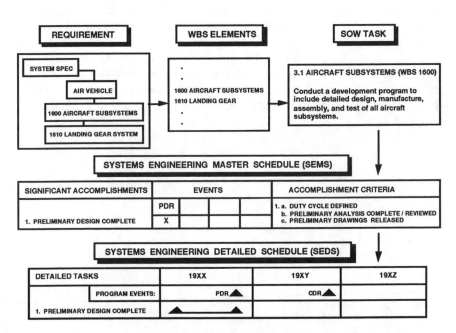

Figure 4-4   *Systems Engineering Master Schedule (SEMS) Interrelationships*

4.6 Other Aspects of Systems Engineering

The following subsections present other perspectives of SE as related to levels of architecture and acquisition phases, SE documentation, and verification. With their related figures, the following views are helpful in understanding the framework in which SE works.

4.6.1 Relation to Levels of Architecture and Acquisition Phases

Systems engineering methodology is used iteratively as the level of system definition progresses. This methodology is repeated as the level of architecture progresses from the overall system level down to the unit level, and is integrated back up into a system architecture. At the system, segment and subsegment levels, design engineering supports the systems engineering effort. At the lower levels of iteration, design engineering assumes responsibility, with systems engineering providing oversight and audits. This SE methodology is used throughout the phases of the acquisition life cycle.

This methodology is applied to levels of architecture and acquisition phases as depicted in Figure 4-5. This figure represents the "waterfall" model of SE. This approach is not recommended for most system development efforts since it leads to unnecessary risk and delays. However, it might be good for development of highly precedented systems, those that are very similar to other systems done in the past (i.e., no new technologies, no new environments, no inexperienced engineers). Other life cycle models are described in Appendix E.

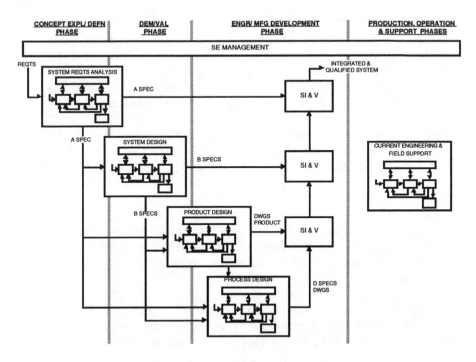

Figure 4-5 *SE Process as Applied to Levels of Architecture and Acquisition Phases*

4.6.2 SE Process in Terms of SE Documentation

Figure 4-6 depicts a notional[30] SE process documentation flow, beginning with the customer requirements.[31] In this particular example, SE performs requirements analysis, then uses Functional Flow Block Diagrams (FFBD) to identify the functional requirements for a successful system that will satisfy customer requirements. The functions are decomposed to the level needed to further define "what" must be done. Using time line analysis, concept of operations, and the customer's requirements of cost, reliability, safety, etc., allocated budgets are developed which are associated with performance allocations, permitting subsystem and design engineers to conduct trade studies to identify feasible "how to" solutions that meet allocated requirements and constraints. The results of the trade studies permit writing a form, fit, and function subsystem or component specification along with identifying the required facilities, equipment, personnel and training which will be further optimized in more detailed iterations of the process. Schematic block diagrams represent the system interfaces with subsystems and major components. Test requirements, peculiarities of the manufacturing process, risk areas, and support issues will all be documented during the functional decomposition, trade studies, detailed design, and support concept selection.

---

[30]As discussed in Chapter 3, there are many *methods* for implementing a *process*. The particular documents and techniques shown here are merely one way of performing systems engineering.

[31]The customer does not always document his requirements. In this case, Systems Engineering should work with marketing and sometimes directly with the customer to define his needs and requirements. Quality Function Deployment (QFD) is an excellent method for doing this.

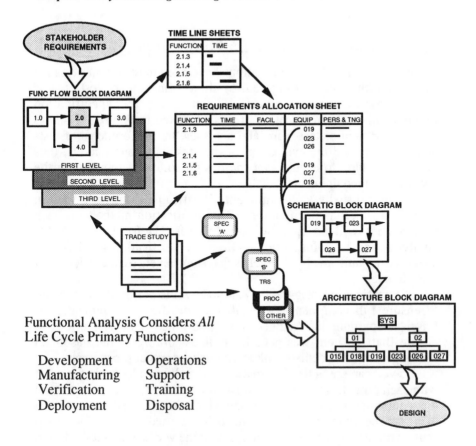

Functional Analysis Considers *All* Life Cycle Primary Functions:

Development	Operations
Manufacturing	Support
Verification	Training
Deployment	Disposal

Figure 4-6 *Relationships Between SE Documentation*

4.6.3 Closing the Loop With Verification

Once the "A" and "B" specifications have been defined and design is completed (the output of Figure 4-6), the verification and test process determines whether the actual design meets the specification requirements. The verification process depicted in Figure 4-7 shows the development of test plans and analyses, the performance of tests and analyses, and the evaluation of the results through audits, thus closing the loop to ensure that the design solution conforms to the specification requirements.

There is usually a Requirements Verification Matrix (RVM) in Section 4 of each specification. The verification method for each requirement will be specified—T for Test, D for Demo, A for Analysis, and I for Inspection. Often the verification level is also specified as unit level, CI level, subsystem level, segment level, or system level.

A Functional Configuration Audit (FCA) is performed at the end to ensure full compliance with all the specified requirements. A Physical Configuration Audit (PCA) is performed on the test articles to ensure that the item was actually built according to the baselined drawings, source code lists, link instructions, or other procedures.

As you can see in the figure, there are really two loops that interact with each other to give you full verification. The first loop starts with the requirements that lead to a design which, in turn, leads to some sort of prototype or model of the item to be verified. The second loop starts with the requirements which lead to the verification methods and procedures for checking compliance to these requirements.

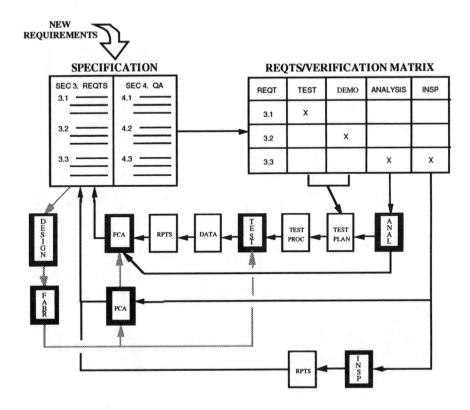

Figure 4-7 *Closing the Loop for Requirements With Prototype Fabrication and the Requirements Verification Process*

chapter five

Systems engineering process details

"Where there is no vision, the people perish."
— Proverbs 29.18

Systems engineering is partly about providing a vision to the project of the ultimate goal for a system. That vision must be concerned with satisfaction of the end users. You might comply legally with all the requirements, but if the end user is unhappy then you have failed.

The systems engineering process is depicted in Figure 4-2 (repeated below in Figure 5-1). Chapters 6 through 8 include detailed process diagrams, tables of inputs and outputs for each activity, and task descriptions which detail different aspects of the SE process. The tasks for each activity are tagged with a task number which can be used to facilitate process tailoring and project planning, to map the tasks to a Work Breakdown Structure, to correlate with SE methods and tools, or to perform a skills inventory.

5.1 Process Tailoring

Chapter 9 gives guidance on tailoring this process. A tailoring worksheet is provided in that chapter to assist in the tailoring task. Also, the input/output tables provided in Chapters 6 through 8 can be used as "checklists" for process tailoring or project planning.

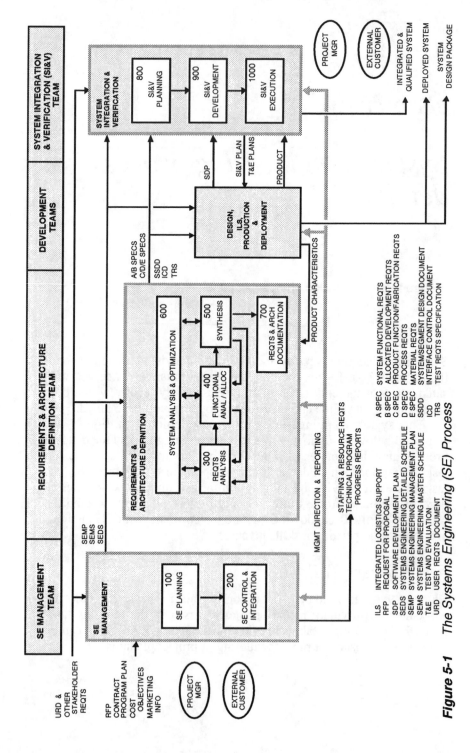

Figure 5-1 *The Systems Engineering (SE) Process*

5.2 Graphical Notation

The symbols used in the detailed task flowcharts in the following Chapters are defined in Figure 5-2. The lines in these diagrams represent task execution, not data flow. The actual sequence of task execution and number of iterations will be tailored for the project. Data flow is shown in the input/output tables. The reviews (circle R symbol) are internal management reviews. The peer inspections (circle I symbol) represent opportunities for applying the In-Process Quality Inspection (IPQI) process per Task 211 (see Appendix G).

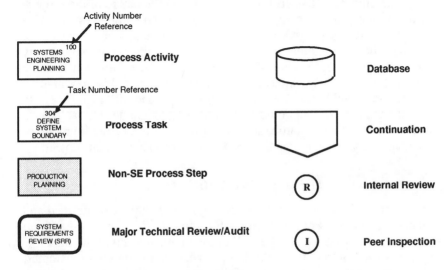

Figure 5-2 *Process Diagram Legend*

5.3 Process Inputs and Outputs

The SE Process has several inputs. The User Requirements Document (URD) contains the needs, hopes, desires, expectations, etc. from the end user's perspective. The URD also describes the different types of users, special characteristics and limitations of each user type, and expected usage scenarios for the system or product under development. The URD "requirements" are often not stated in technical terms. The objective of the Requirements and Architecture Definition subprocess is to translate these into technical requirements for the development teams.

Other stakeholder requirements may drive the design decisions. These requirements could come from various sources: manufacturing, corporate policy, company standards, public law, government regulation, community building code, test engineering, transportation facilities, etc.

There are various other drivers to the SE Process. A Request for Proposal (RFP) may be provided as part of a solicitation in the government acquisition process. There may be a contract associated with another company or an acquisition agency of the government. The project may be part of a larger program which has its own program plan which must be followed. Marketing may provide some cost objectives and other such information as competitive data, technical specs from similar products, etc.

The outputs of the SE Process depend on the nature of the project. An output could be a completely integrated and qualified system. This could be used at a higher level to be integrated into a larger system or it could be for customer or user acceptance testing. The project may involve actual deployment of a system. In this case, the output is the completely (or perhaps partially) deployed system. This may involve some sort of field acceptance testing prior to "hand-off" to the user. In other cases, the output could be merely a technical design package to be used by other organizations for production and deployment or perhaps for your own organization to use later in a different project.

5.4 Activity Inputs and Outputs

For each activity of the SE Process, inputs and outputs are defined and listed in tables. For each input, the table defines where that item comes from (supplier), the name[32] of the input, and entry criteria.

For each output, the table defines where that item goes (customer, internal or external), the name of the output, exit criteria, and the task from which that output is generated. Also, included in both types of tables is a blank column called "status" that makes these tables useful as checklists during process tailoring and project planning.

---

[32]The names of the documents are intended to be generic. Each organization will need to determine the particular names of these documents for their own use. Often the names of documents will depend on which particular customer that project is dealing with. Hence, flexibility of nomenclature is essential.

Also, there is no intention of requiring that any of the input or output documents be treated as separate documents. Often, these documents are combined and it usually is a good idea to combine them where feasible since this will likely save money.

chapter six

Systems engineering management subprocess

"Plans are nothing, planning is everything."
—General Dwight D. Eisenhower

The purpose of the SE Management Subprocess is to ensure that the development tasks are accomplished efficiently and effectively through proper planning and control of the SE process and development team efforts. SE Management Subprocess should ensure the proper balance of quality, cost, schedule, risk and performance. Figure 6-1 shows the major activities of this subprocess.

SE Planning leads the systems engineering effort with the tailored integration and coordination of all planning activities. The Systems Engineering Management Plan (SEMP) is the concise technical management plan for the integration of all engineering activities. Execution of the SEMP and the specialty engineering project plans is then monitored, managed, and audited by SE Management continually throughout the project.

SE Control and Integration includes several *key tasks* that are incorporated into the SE process during initial planning. These tasks include:

- Design-To-Cost (DTC) Monitoring

- Risk Management

- Technical Performance Measurement (TPM)

- Configuration Management (CM)

- Data Management (DM)

- Technical Reviews and Audits

- Trade Studies and Effectiveness Analyses

- Life Cycle Cost (LCC) Analysis

- Technical Discipline Integration

    Comprehensive planning, careful management, and coordinated integration of the myriad of disciplines, functions, specialties, and processes, within ever present "real world" constraints, will lead to a higher probability of successfully meeting the real user and customer needs.

    Iteration will occur between the Planning activity and the Control and Integration activity as the need for replanning arises. These two activities are analogous to the phrase, "Plan the work, then work the plan." They are also consistent with the PDCA approach: Plan, Do, Check, and Act.

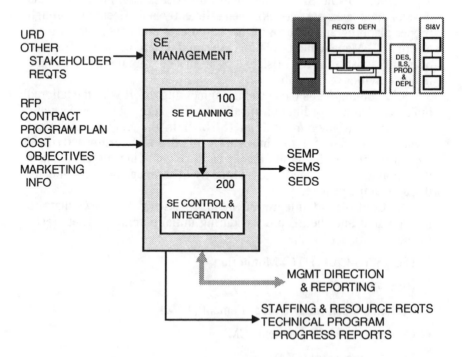

Figure 6-1   SE Management Subprocess

6.1 SE Planning Activity

The SE Planning effort[33] determines how the systems engineering process is to be accomplished and controlled to meet project objectives. Tailoring of the SE process to a specific project depends on the scope, complexity, and phase of the project. This tailoring is documented in the Systems Engineering Management Plan (SEMP) which provides a road map of how the required technical efforts will be managed and conducted over the life cycle of the system products and processes. Figure 6-2 shows the SE Planning tasks.

Task 101 Identify Corporate Resources

Early in the overall project planning stage, a project Systems Engineering Manager (SEM) is designated with the responsibility of forming Integrated Product Teams (IPTs).

It is imperative that representatives from all disciplines expected to be involved in the project participate at the earliest practical stages. This early and effective teaming is at the heart of successful integrated product and process development. Initial efforts will focus on selection of persons with the necessary expertise for the multi-disciplinary project team.

This task also involves identification (and obtaining commitments for project use, if necessary) of development facilities, equipment and technology. This may include patent rights, copyrights, teaming arrangements, notification to use,[34] and non-disclosure agreements.

Task 102 Tailor SE Process

Tailoring the SE process to a specific project depends on the scope, complexity, and phase of the project. Generally, tailoring will define the scope of the effort. Concurrently, related processes such as Risk Management and Configuration Management are tailored for incorporation into the overall systems engineering process. Tailoring Guidance is provided in Chapter 9.

---

[33]It is assumed that the SE Management team is formed by the Project Manager before the SE Planning effort begins.

[34]A notification to use (NTU) action is intended to inform vendors and suppliers that you intend to use one or more of their products. This will allow this vendor or supplier to notify you when they have plans on changing or discontinuing that product. This approach greatly reduces risk by eliminating some of the surprises that occur when you "design-in" someone else's component only later to find out that it may no longer be available, or perhaps it was only available in small quantities.

Task 103 Define Methodologies for System Analysis, Optimization & Development

Determine the scope and procedures for application of Risk Management, Measures of Effectiveness (MOEs) and trade studies. Define any special techniques, procedures and tools needed for effectiveness analyses. Define plans for simulations and modeling.

Task 104 Develop Technology Insertion Approach

A plan will be developed for appropriate level and timing of technology insertion. This may include alternate approaches, Preplanned Product Improvement (P3I) approaches, or other strategies to take advantage of new technologies to meet systems needs. An initial technology assessment will be performed to identify technology constraints applicable to the Requirements and Architecture Definition Subprocess.

Task 105 Identify Corporate Procedures & Planning Baselines

Determine applicable policies and procedures for the project. Identify related projects in your organization working on similar problems. Identify marketing strategies and plans that affect your project or the market segment for your product.

Task 106 Assess Technical Program Risks and Issues

Receive initial risk assessment from risk process to identify technical risk areas for systems engineering management emphasis. Mitigation plans must be developed to track the identified high-level risks.

Task 107 Define Technical Performance Measurement (TPM) Parameters & Procedures

On the basis of technical risks and priorities, determine parameters that will receive management focus and will be tracked during the project to assess conformance with specification and contract requirements. These Technical Parameters (TPs) must be related to the Measures of Effectiveness (MOEs) and Measures of Performance (MOPs).

Task 108 Define Technical Review and Audit Plans

Define the purpose, scope, number, timing, and formality of internal and external technical reviews and audits. This will be based on the scope of the project and contract requirements.

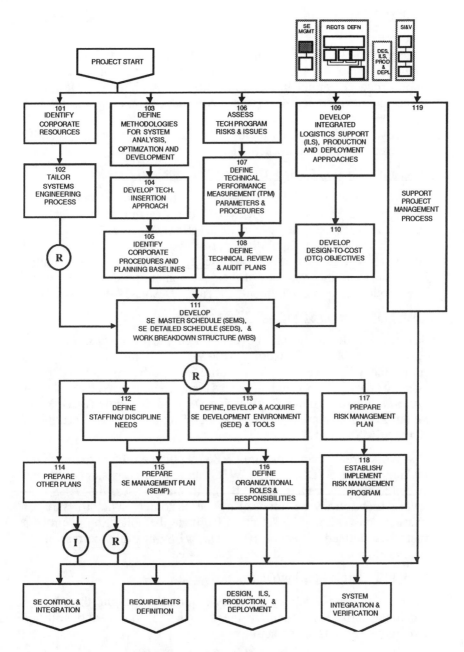

Figure 6-2 *SE Planning Activity*

Task 109 Develop Integrated Logistics Support (ILS), Production, and Deployment Approaches

Determine how logistics, production, and deployment activities will be integrated into the project. This includes active participation by appropriate personnel in Requirements and Architecture Definition teams and development teams (see Appendix B on multi-disciplinary team concept).

   Develop release plans for the system and its products. Identify features to be provided in each release.

Task 110 Develop Design-To-Cost (DTC) Objectives

Rigorous cost goals should be established prior to and during development. The control of system costs (acquisition, operating, support and disposal) to these goals is achieved by practical trade-offs between operational capability, performance, life cycle costs, and schedule. Cost, as a key design parameter, is addressed on a continuing basis and as an inherent part of the development processes of design, ILS, production and deployment. Prepare a Design-To-Cost (DTC).

Task 111 Develop SEMS, SEDS, and WBS

Technical milestones will be defined in the *event-based* Systems Engineering Master Schedule (SEMS). Each major activity and milestone will have completion criteria specifically defined (e.g., specific analysis completed, drawings completed, prototype results). A *calendar-based* Systems Engineering Detailed Schedule (SEDS) will be developed to support the SEMS. See Section 4.5 for a description of the SEMS and SEDS.

The Work Breakdown Structure (WBS) should be refined to match the tasking in the SEDS.[35] The WBS should be consistent with the Architecture Block Diagram (ABD). In many cases, the WBS top-level structure may be imposed by the customer. The structure for charging costs should be set up to facilitate the collection of process metrics as defined in Section 10.3. The WBS must fully correspond to the system architecture and specification tree.

Task 112 Define Staffing/Discipline Needs

Estimate staffing levels by category required to execute the planned technical activities. Identify the knowledge, skills, and abilities required for each task and activity.

---

[35]The WBS will be further refined during the Synthesis steps as the design solution evolves. It is important to not constrain the possible system solutions because of an inflexible WBS.

Task 113 Define, Develop & Acquire Systems Engineering Development Environment (SEDE) and Tools

Define the appropriate development environment for the project, including automation needs. If necessary, develop and/or acquire the tools and facilities for all disciplines on the project. Define the requirements for an information management system and for using existing information management system elements. Define and plan for the training needed for using the SEDE. (Training and tools are described in Chapter 10.)

Task 114 Prepare Other Plans

Prepare and coordinate the development of other plans for the project that may be required, especially by contract, such as specialty engineering plans, ILS plans, etc.

Task 115 Prepare Systems Engineering Management Plan (SEMP)

The results of previous tasks will be documented in a SEMP which will be fully coordinated within the project and within company or agency, as appropriate. See Section 4.4 for a description of a SEMP.

Task 116 Define Organizational Roles and Responsibilities

Create teams that will develop the products and processes with an integrated approach for meeting all requirements. See Chapter 11 and Appendix B on the multi-disciplinary team concept.

These teams will consist of Integrated Product Teams (IPTs) and other ad hoc teams. A hierarchy of teams ("team of teams") should be used that matches the system architecture. Some of these teams will be short-term and tasked with very specific objectives (e.g., to identify and define external interfaces). Normally there will be one IPT for each Configuration Item (CI) or set of related CIs.

The WBS elements need to be completely mapped to all IPTs. Ideally, this should be a one-to-one mapping between WBS elements and IPTs such that there are no WBS elements "shared" between teams. Otherwise, there would not be clear accountability.

Identify a Peer Inspection Coordinator. This person will be responsible for implementation of the In-Process Quality Inspection (IPQI) process. This person will develop project-specific inspection checklists and collect peer inspection metrics. (Peer inspection metrics are a crucial means for determining the root cause of development problems and can lead to significant process improvements. See Appendix G for more information on these IPQI metrics.)

Identify the Chief Systems Engineer (CSE) for the system to be developed. The CSE will be the lead engineer and chief architect for the entire system. This may be the same person as the Systems Engineering Manager (SEM).

Task 117 Prepare Risk Management Plan

Define the appropriate Risk Management procedures and templates. Document these in a project Risk Management Plan.

Task 118 Establish/Implement Risk Management Program

Initiate Risk Management in accordance with the risk process as tailored and documented in the Risk Management Plan.

Task 119 Support Project Management Process

Provide an interface between the technical aspects of the project and the overall project management process during the SE Planning activities to coordinate technical efforts with the overall project. This also includes the technical interactions with the external customers and subcontractors.

Table 6-1A   SE Planning Input Checklist

Supplier	Input	Entry Criteria	Status
Functional Mgmt	Planning Team Identified	Agreement by Team	
PM	Request for Proposal (RFP)	Appvd for Use by PM	
PM	Proposal	Appvd for Use by PM	
PM	Marketing Report	Appvd for Use by PM	
PM	Business Plans	Appvd for Use by PM	
PM	Policies and Procedures	Appvd for Use by PM	
PM	Contract	Appvd for Use by PM	
PM	Statement of Work (SOW)	Appvd for Use by PM	
PM	Contract Data Requirements List (CDRL) or Subcontract Data Requirements List (SDRL)	Appvd for Use by PM	
PM	User Requirements Document (URD)	Appvd for Use by PM	
PM	Technical Specifications	Appvd for Use by PM	
PM	Engineering Budget	Appvd for Use by PM	
PM	Project Management Plan	Appvd for Use by PM	
PM	Project Master Schedule	Appvd for Use by PM	
PM	Contract Work Breakdown Structure (WBS)	Appvd for Use by PM	
PM	Organizational Breakdown Structure (OBS)	Appvd for Use by PM	
PM	Resource Planning	Appvd for Use by PM	
PM	Task Definition	Appvd for Use by PM	
PM	Project Modification Requests	Appvd for Use by PM	
Data Mgmt	Standards and Specifications	Appvd for Use by SEM	
Data Mgmt	Data Item Descriptions (DIDs)	Appvd for Use by SEM	
T&E	Test and Evaluation (T&E) Capabilities and Limitations	Appvd by PM and T&E Mgmt	
Production	Production Capabilities and Limitations	Appvd by PM and Prod Mgmt	
Deployment	Deployment Capabilities and Limitations	Appvd by PM and Depl Mgmt	
Logistics	Logistics Capabilities and Limitations	Appvd by PM and ILS Mgmt	

Table 6-1B   SE Planning Output Checklist

Customer (1)	Output	Exit Criteria	Task #	Status
Systems Engrs, SE PMT	Tailored SE Process Document	Insp & Appvd by SEM	102	
ALL	Effectiveness and Trade Studies Plan	Insp & Appvd by SEM	103	
ALL	Simulation/Modeling Plan	Insp & Appvd by SEM	103	
ALL, Func Mgmt	Technology Insertion Plan	Insp & Appvd by SEM	104	
ALL	Risk Assessment Report	Approved by SEM	106	
ALL	TPM Charts	Insp & Appvd by SEM	107	
ALL	Technical Review Plan	Insp & Appvd by SEM	108	
ALL	Release Plan	Insp & Appvd by SEM	109	
ALL, Func Mgmt	Design-To-Cost Plan	Insp & Appvd by SEM	110	
ALL, Func Mgmt	Design-To-Cost Allocations	Insp & Appvd by SEM	110	
ALL	Systems Engineering Master Schedule (SEMS)	Insp & Appvd by SEM	111	
ALL	Systems Engineering Detailed Schedule (SEDS)	Insp & Appvd by SEM	111	
ALL	Work Breakdown Structure (WBS)	Insp & Appvd by SEM	111	
PM	Staffing/Resource Request	Approved by SEM	112	
ALL, Func Mgmt	Systems Engineering Development Environment (SEDE) Plan	Insp & Appvd by SEM	113	
ALL, Func Mgmt	Systems Engineering Tools Plan	Insp & Appvd by SEM	113	
ALL	Configuration Management Plan	Insp & Appvd by SEM	114	
ALL	Data Management Plan	Insp & Appvd by SEM	114	
ALL, SpE Plng Proc	Specialty Engineering Plans	Insp & Appvd by SEM	114	
ALL, SpE Plng Proc	ILS Plan	Insp & Appvd by SEM	114	
ALL, SpE Plng Proc	Producibility Plan	Insp & Appvd by SEM	114	
ALL, SpE Plng Proc	Deployability Plan	Insp & Appvd by SEM	114	
ALL	Other Plans	Insp & Appvd by SEM	114	

Chapter 6: Systems Engineering Management

ALL (1)	Systems Engineering Management Plan (SEMP)	Insp & Appvd by SEM	115	
ALL	Integrated Product/Process Teaming Plan	Insp & Appvd by SEM	116	
ALL, Func Mgmt	Technical Integration Teaming Plan	Insp & Appvd by SEM	116	
ALL	Risk Management Plan	Insp & Appvd by SEM	117	
ALL	Draft Business Baseline	Insp & Appvd by SEM	119	

(1) "ALL" customers indicates PM, SE Control & Integration, Requirements and Architecture Definition, SI&V, and Development Teams.

6.2 SE Control and Integration Activity

"There go the people. I must follow them, for I am their leader."

—Alexandre Ledru-Rollin

SE Control and Integration includes the management tasks of the SE process to guide, coordinate, focus, and balance the other SE tasks as well as the design, manufacturing, deployment, and ILS processes. Included are risk management, technical program progress assessments, documentation control, and integration of various efforts.

The SE Control and Integration Activity (see Figure 6-3) oversees the tailored SE approach as documented in the SEMP. The purpose of SE Control and Integration is to ensure that the SE tasks are accomplished efficiently and effectively through proper planning and control of the SE process and by appropriate development teaming.

SE Control manages the technical program and provides continuous support to design, production and deployment planning and execution. Design-To-Cost (DTC) objectives are assessed, reviewed and managed. Risks of the technical program are assessed, reviewed and managed. Traceability of the technical baseline and any changes will be maintained while documenting the configuration, technical data, trade studies and effectiveness analyses.

SE Integration involves the coordination of diverse engineering specialties, product development teams, etc. See Appendix B on the multi-disciplinary team concept.

SE Control and Integration includes several key tasks that are incorporated into the SE process during initial planning. These tasks include:

- Risk Management Process

- Technical Performance Measurement (TPM) Process, which addresses updating of TPMs based on changes in risk, new parameters or deletions

- Configuration Management (CM) Process

- Data Management (DM) Process

- Technical Reviews and Audits Process.

Task 201 Project Engineering

This task includes the management efforts of the Systems Engineering Manager (SEM) to maintain the correct staffing, equipment and facility levels, manage cost and schedule for the SE tasks, and coordinate among the process participants and the project manager to execute the SE tasks.

Project Engineering includes leading continuous process improvement efforts, leading Integrated Product Teams (IPTs), coordinating all engineering control activities, and replanning development activities, as required. It also includes technical interactions with the external customer. The SEMP and SEMS will be updated as required.

Establish and maintain engineering policies and procedures for the project. Coordinate training of project personnel and teams in these project-specific and other policies and procedures.

The SEM will coordinate and review process metrics from all functional areas and report these to management. Document the lessons learned for use by other projects. Implement lessons learned from other projects.

Task 202 Risk Management

The Risk Management Process provides a systematic approach for the identification of risk and potential sources of risk. It will identify actions, events or products that have a negative impact on achieving the goals or objectives of the project and will develop actions to reduce the probability or consequences of those risk items.

This task will identify risk drivers, sensitivity of interrelated risks, and impact on related technical efforts. Identification and monitoring of technical risks may use as a guide DOD 4245.7M, "Transition from Development to Production." Risk assessments should include consideration of subcontractor and vendor quality.

Task 203 Technical Parameter (TP) Tracking

Technical parameter tracking is part of Technical Performance Measurement (TPM). TPM updates the status of parameters based on technical progress or changes in parameters, limits, or objectives that are in turn based on changes in risk, requirement changes, or the addition or deletion of parameters.

Task 204 Assess and Review Technical Program Progress

Evaluate technical progress of the entire project based on TPM results and SEMS/SEDS criteria, and review results with the Project Manager and other interested parties.

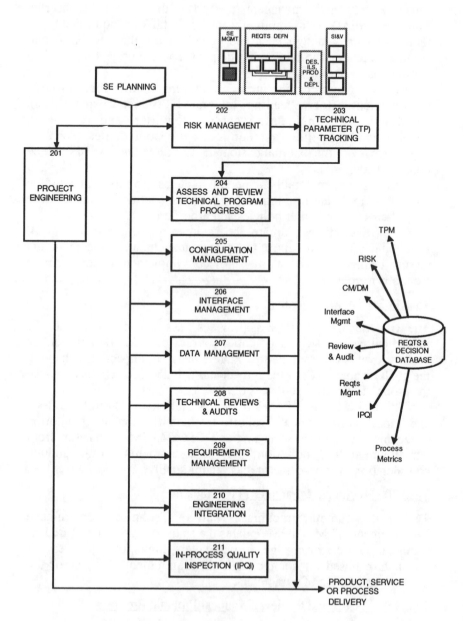

Figure 6-3 *SE Control and Integration Activity*

Task 205 Configuration Management

Configuration Management (CM) includes the functions of identification of products and processes to be controlled (through specifications, ICDs, drawings, etc.), change control (ECPs, etc.), and status accounting. CM should be in accordance with EIA Standard-649 or other appropriate company or industry procedures.

This task includes establishing and maintaining the Configuration Control Board (CCB). The Systems Engineering Manager (SEM) will usually chair the CCB.

Task 206 Interface Management

Interface Management includes interface definition, interface control, interface compatibility assessment, and interface coordination through an Interface Control Working Group (ICWG). This task includes managing interfaces with prime contractors and subcontractors.

Task 207 Data Management

Data Management (DM) supports the development, control, and delivery of required technical data throughout the project. It also supports information management which involves the use of computer networks, terminals, databases, libraries, communication links, and other shared resources.

Task 208 Technical Reviews and Audits

Systems engineering is responsible for the conduct of several reviews and audits during execution of the design, production and deployment processes. Required reviews and audits are scheduled in the SEDS and occur at specified points for determining design completeness, traceability, and ability to meet requirements. The exit criteria for each review and audit should be specified in the SEMS.

Each review or audit has a specific objective as a milestone in the design process. The objectives become increasingly more detailed and definitive as development progresses. Appendix F describes reviews and audits (for DOD acquisitions) and their relationship to the development activities.

Essential elements of this task are as follows:

- Ensure that SEMS accomplishment criteria have been satisfied.

- Establish review philosophy and procedures.

- Ensure adequate internal reviews are being planned and executed properly.

- Establish requirements and place under CM control.

- Ensure that reviews cover all related engineering specialty disciplines.

- Demonstrate that required relationships, interactions and interfaces among items, functions, subsystems and configuration items have been addressed.

- Ensure that requirements have been decomposed and allocated appropriately, and traceability has been maintained.

- Identify action items to be completed to continue development.

- Assess risks for the item being reviewed and for the system as a whole.

- Ensure that supportability, producibility, testability, and deployability are adequately addressed.

Task 209 Requirements Management

Establish a historical database of technical decisions and requirements for future reference. The database will be the primary means for maintaining requirements traceability. All product and process requirements should be maintained in this database.

Develop a Requirements Traceability Matrix (RTM) as a report from the database. The RTM will map the requirements to subsystems, Configuration Items (CIs), and functional areas. The RTM should be reissued on a regular basis to communicate the latest requirements and allocations.

Identify technical budgets that need to be tracked. Develop a philosophy and approach for managing the technical margins. Reallocate margin as appropriate according to risk assessments.

Ensure that the allocated requirements and the results from the synthesis tasks are consistent and traceable to the work packages in the WBS. Refine the WBS as required. Ensure that consistency is maintained between the Statement of Work (SOW), the Organizational Breakdown Structure (OBS), and the Architecture Block Diagram (ABD). Also, ensure that the cost objectives are being met in accordance with the Cost Breakdown Structure (CBS).

Task 210 Engineering Integration

Provide coordination of diverse technical disciplines and integration of the development project tasks. This includes use of integrated product development approaches and product development teams. See Appendix B on the multi-disciplinary team concept. Ensure that specialty engineering[36] is properly represented on all technical teams, and define the scope and timing of the specialty engineering tasks.

Establish and maintain a System Integration Working Group (SIWG) and Failure Review Board (FRB). Coordinate development of all technical plans.

Task 211 In-Process Quality Inspection (IPQI)

Perform IPQI on all documents as indicated in the output checklist tables in this book. Basically, the IPQI process involves an inspection of a document or a physical item against its requirements. Usually an IPQI is performed by peers. The major defects discovered during an IPQI may be used to perform root cause analysis to identify opportunities for process improvement. A process and associated metrics for IPQI are described in Appendix G.

---

[36]Specialty engineering involves such disciplines as reliability, maintainability, safety, supportability, operations research, etc.

Table 6-2A   SE Control and Integration Input Checklist

Supplier	Input	Entry Criteria	Status
PM	Marketing Report	None	
PM	Requirements Changes	Appvd by Contracting	
PM	Contract Modifications	Appvd by Contracting	
PM	Approved Engineering Change Proposals (ECP)	Appvd by Ext Customer CCB	
PM	Project Plan Changes	None	
SE Plng	Systems Engineering Management Plan (SEMP)	Appvd by SEM	
SE Plng	Systems Engineering Master Schedule (SEMS)	Appvd by SEM	
SE Plng	Systems Engineering Detailed Schedule (SEDS)	Insp & Appvd by SEM	
SE Plng	Design-To-Cost Plan	Insp & Appvd by SEM	
SE Plng	Design-To-Cost Allocations	Insp & Appvd by SEM	
SE Plng	Systems Engineering Development Environment (SEDE) Plan	Insp & Appvd by SEM	
SE Plng	Tailored SE Process Document	Insp & Appvd by SEM	
SE Plng	Effectiveness and Trade Studies Plan	Insp & Appvd by SEM	
SE Plng	TPM Charts	Insp & Appvd by SEM	
SE Plng	Technical Review Plan	Insp & Appvd by SEM	
SE Plng	Technology Insertion Plan	Insp & Appvd by SEM	
SE Plng	Technical Integration Teaming Plan	Insp & Appvd by SEM	
SE Plng	Configuration Management Plan	Insp & Appvd by SEM	
SE Plng	Specialty Engineering Plans	Insp & Appvd by SEM	
SE Plng	ILS Plan	Insp & Appvd by SEM	
SE Plng	Producibility Plan	Insp & Appvd by SEM	
SE Plng	Deployability Plan	Insp & Appvd by SEM	
SE Plng	Risk Management Plan	Insp & Appvd by SEM	
SE Plng	Risk Assessment Report	Appvd by SEM	
SE Plng	Other Plans	Insp & Appvd by SEM	
CAM	Metrics	Appvd by CAM	
CAM	Cost/Schedule Variance Reports	Appvd by CAM	
CAM	CDRL/SDRL Status	Appvd by CAM	
CAM	Risk Analysis Reports	Appvd by CAM	
CAM	Action Item Status	Appvd by CAM	
CAM	TPM Status	Appvd by CAM	
CAM	SEMP Variance Report	Appvd by CAM	
CAM	Change Request (CR)/Change Order (CO)	Appvd by CAM	
CAM	Technical Data Items	Insp & Appvd by CAM	
ALL	Jeopardy Reports	None	

Table 6-2B SE Control and Integration Output Checklist

Customer	Output	Exit Criteria	Task	Status
ALL*	Systems Engineering Policy	Appvd by SEM	201	
ALL	Systems Engineering Master Schedule (SEMS) Updates	Appvd by SEM	201	
ALL	Systems Engineering Detailed Schedule (SEDS) Updates	Insp & Appvd by SEM	201	
ALL	Tailored SE Process Document Updates	Insp & Appvd by SEM	201	
ALL	Technical Review Plan Updates	Insp & Appvd by SEM	201	
ALL	Other Plan Updates	Insp & Appvd by SEM	201	
PM	Staffing/Resource Request	Appvd by SEM	201	
PM, SE PMT	Lessons Learned Report	Insp & Appvd by SEM	201	
SE PMT	SE Process Metrics Data Forms	Insp & Appvd by SEM	201	
Various PMTs	Metrics Data Forms for Other Functional Processes	Insp & Appvd by SEM	201	
PM	Technical Data Items	Appvd by SEM	201	
PM	Risk Assessment Report	Appvd by SEM	202	
ALL	TPM Chart Updates	Insp & Appvd by SEM	203	
PM	Trade Study Reports	Insp & Appvd by SEM	204	
PM	Technical Program Progress Report	Appvd by SEM	204	
PM	Change Status Report (CSR)	Appvd by CM	205	
PM	Engineering Change Notice (ECN)	Appvd by CM	205	
PM	Configuration Status Accounting Report (CSAR)	Appvd by CM	205	
External Customer	Engineering Change Proposals (ECP)	Appvd by Proj CCB	205	
ALL	Approved ECPs	Appvd for Use by SEM	205	
PM	Technical Audit Reports	Appvd by SEM	208	
ALL	Approved Specifications	Appvd for Use by SEM	208	
ALL	Requirement Status Report	Appvd by SEM	209	
ALL	Requirements Traceability Matrix (RTM)	Appvd by SEM	209	
ALL	TPM Status Report	Appvd by SEM	209	
ALL	Risk Status Report	Appvd by SEM	209	
ALL	Peer Inspection Status Report	Appvd by SEM	211	
SE PMT	Peer Inspection Metrics	Appvd by SEM	211	

* "ALL" customers indicates PM, Requirements and Architecture Definition, SI&V, and Development Teams.

chapter seven

Requirements and architecture definition subprocess

"Architecture is frozen music."

—Madame de Stael

"Necessity never made a good bargain."

—Benjamin Franklin

The Requirements and Architecture Definition Subprocess (see Figure 7-1) provides an orderly and iterative definition of the problem and development of the solution. Requirements Analysis defines the boundary of the problem and parameters to be satisfied. Functional Analysis describes the intended behavior of the system in its environment. The "problem domain" will be defined by these requirements and functions. The "solution domain" will be defined during the Synthesis tasks.

The "verification loop" from Synthesis to Requirements Analysis ensures that the solution domain maps correctly to the problem domain. System Analysis and Optimization analyzes the alternative solutions for their effectiveness and narrows the choices for further development. The requirements for the final choices will be documented in specifications and interface documents during the Requirements and Architecture Documentation task.

The requirements will be defined for both the system products and the related processes, such as manufacturing, verification, deployment, support, and disposal.

The feedback loop from Functional to Requirements Analysis indicates the iterative and interrelated nature of these tasks in that as

117

new functions are identified, new derived requirements will need to be defined to quantify the functionality. Synthesis defines potential architecture alternatives or product/process designs that will meet the requirements. During successive iterations of the process, one or more of the design concepts will be synthesized for each system concept.

This subprocess requires technology assessment, projection and selection. This feeds into risk assessment and management. Technology constraints also flow into producibility and supportability.

During each project phase, design develops still greater levels of detail until drawings and procurement specifications are produced. The feedback loop between Synthesis and Functional Analysis ensures that as design decisions are made, specific functions, particularly at the lower levels, will be added or rearranged. The feedback to requirements indicates the need to confirm (or verify) that proposed solutions meet the requirements.

The systems engineering process ensures that all technical requirements are consistent with and traceable to higher level requirements. Systems engineering controls and audits all design documentation.

Specification documents, architecture and design documentation, interface definitions, and test requirements are given to the development teams as an input to their activities. These development teams will provide product characteristics back to the Requirements and Architecture Definition team for use in architectural trade studies and system effectiveness analyses.

The Requirements and Architecture Definition Team that performs this task should consist of all disciplines that will add value to the process. This must include persons from design, manufacturing, QA, procurement, deployment, test, etc.

7.1 Requirements Analysis Activity

System level requirements will be developed using customer requirements in the RFP, contract, Statement of Work (SOW), CDRL/SDRL, User Requirements Document (URD), technical specification, or other documents (see Table 7-1A). These requirements include production, deployment, test and ILS requirements, as well as operational performance. Development team members will provide additional sources of derived requirements.

A requirements/decision database is established and maintained. The system mission profile is either provided by the customer or developed by the project. The operational and support concepts are refined or defined, and documented in an Operational Concept Document (OCD).

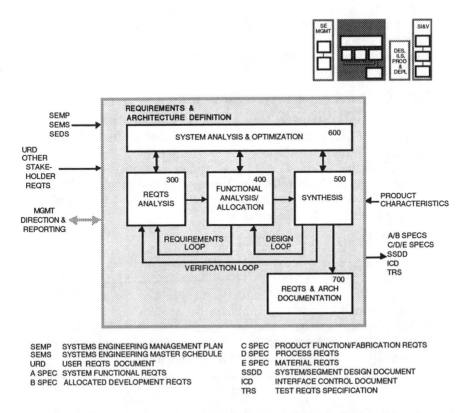

SEMP   SYSTEMS ENGINEERING MANAGEMENT PLAN   C SPEC PRODUCT FUNCTION/FABRICATION REQTS
SEMS   SYSTEMS ENGINEERING MASTER SCHEDULE   D SPEC PROCESS REQTS
URD   USER REQTS DOCUMENT   E SPEC MATERIAL REQTS
A SPEC SYSTEM FUNCTIONAL REQTS   SSDD   SYSTEM/SEGMENT DESIGN DOCUMENT
B SPEC ALLOCATED DEVELOPMENT REQTS   ICD   INTERFACE CONTROL DOCUMENT
                                                TRS   TEST REQTS SPECIFICATION

Figure 7-1 *Requirements and Architecture Definition Subprocess*

Requirements are developed concurrently for all functions and subfunctions based on the entire system life cycle. Requirements analysis identifies the criticality and sensitivity of each of these requirements. Figure 7-2 shows the tasks for this activity.

Task 301 Collect Stakeholder Requirements

Identify the stakeholders. Talk to these stakeholders whenever possible. Identify their wants, needs, and expectations. Organize and categorize these requirements. Quantify these requirements indicating their relative importance to one another.

Determine the current level of satisfaction with existing products for each requirement. Relate these to the business goals of your organization and compare them to competing products. Rank order and normalize the relative scores for these requirements.

Define customer and user classes and characteristics of each class. These customers and users will be related to the system primary functions described in Task 308. Quality Function Deployment (QFD) and Analytical Hierarchy Process (AHP) are two methods for performing this task.

Task 302 Define System Mission/Objective

Using the needs defined in Task 301, develop a complete definition of the purpose(s) of the system and its operational profile. Define the threats and available platforms.

Define the targeted market segment. Identify any relevant business case analyses related to this market and product. Perform additional business case analyses to determine the key parameters relevant to each applicable market.

Task 303 Define System Scenarios

Define the expected scenarios of the system. From a black-box perspective, define the stimuli to be encountered and response to each stimulus. Prioritize these scenarios according to the probability of occurrence and severity of strain on the system. The system test philosophy and approach will be based on these system scenarios. Test cases will be developed from these scenarios. Functional failure mode analysis will also be based on these scenarios.

Define the business model that determines the key "success criteria" for the targeted market segments.

Task 304 Define System Boundary

Define the internal and external elements that will be involved in accomplishing the system purpose or mission(s) and the boundaries of the system.

Define the system boundary in terms of both space and time. What are its physical boundaries? What are its operational boundaries? When does the system start performing its mission or objective? When will the system (or its components) be disposed of?

Task 305 Define Environmental & Design Constraints

Identify and document the constraints that will limit or define the system's performance or design, including cost constraints. Design-To-Cost (DTC) goals should be established. The hierarchy of DTC goals will be documented in a Cost Breakdown Structure (CBS).

The design constraints should include such "non-functional" requirements as power, volume, weight, dimensions, etc. The environmental constraints should be defined for all system scenarios (Task 303) and for all primary functions (Task 308).

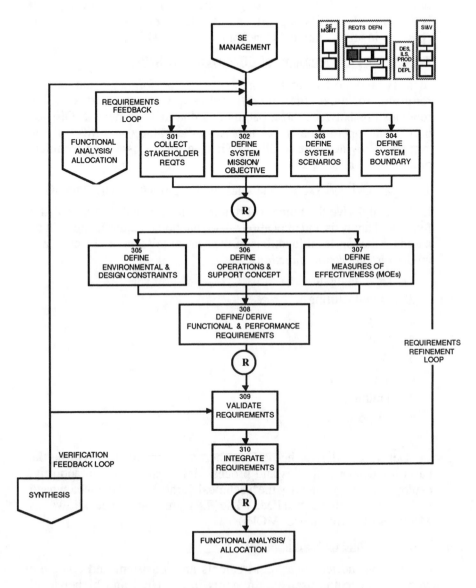

Figure 7-2 *Requirements Analysis Activity*

Task 306 Define Operations & Support Concept

Identify and document the operational and logistics support approaches or constraints that will drive design. The support concept development is part of the Logistics Support Analysis (LSA) process. Document the concepts in an Operational Concept Document (OCD).

Task 307 Define Measures of Effectiveness (MOEs)

Identify and document the most critical performance parameters required to meet the operational requirements and develop relationships between those parameters that drive design. MOEs are used at an operational level to assess the value (or utility) of the system. Provide these MOEs to SE Control for assessing technical program progress.

Task 308 Define/Derive Functional and Performance Requirements

Define and derive the functional requirements that will be the basis for functional analysis and allocation. The functional requirements will be based on the top-level functions of the system mission or objective and on the *system primary functions*:

1) Development
2) Manufacturing
3) Test
4) Deployment
5) Operations
6) Support
7) Training
8) Disposition

   Define and derive the performance requirements that will be the basis for design activities and process development (manufacturing, deployment, etc.). Identify the Technical Parameters (TPs) which need to be tracked through TPM. The TPs are sometimes known as Measures of Performance (MOPs) .

Task 309 Validate Requirements

Ensure that the technical requirements are consistent and complete with respect to the user requirements in the URD and higher-level specifications, and with any other stakeholder requirements.

Task 310 Integrate Requirements

Assure that the requirements trace to the system performance and design. Additionally, make sure all SE allocation budgets are consistent. The allocation budgets should include cost constraints in addition to technical and performance constraints. This activity deals with the problem domain.

Iterate any prior tasks to refine the requirements as appropriate before proceeding to the Functional Analysis/Allocation Activity.

Table 7-1A   *Requirements Analysis Input Checklist*

Supplier	Input	Entry Criteria	Status
PM	Request for Proposal (RFP)	Appvd for Use by PM	
PM	Proposal	Approved for Use by PM	
PM	Contract	Appvd for Use by PM	
PM	Statement of Work (SOW)	Approved for Use by PM	
PM	CDRL/SDRL	Appvd for Use by PM	
PM	User Requirements Document (URD)	Appvd for Use by PM	
PM	Test and Evaluation Master Plan (TEMP)	Appvd for Use by PM	
PM	Technical Specifications	Appvd for Use by PM	
SE Mgmt	TPM Charts	Insp & Appvd by SEM	
SE Mgmt	Approved Engineering Change Proposals (ECP)	Appvd by Cust CCB	
CM	Modification Request (MR), Change Order (CO)	Appvd by Project CCB	
SRR	Preliminary System/Segment Specification (Type A)	Reviewed	
SRR	Preliminary System/Segment Interface Control Document (ICD)	Reviewed	
SRR	Preliminary Operational Concept Document (OCD)	Reviewed	
SRR	System Requirements Review (SRR) Results Documentation	Reviewed	
SDR	System/Segment Specification (Type A)	Reviewed	
SDR	System/Segment Interface Control Document (ICD)	Reviewed	
SDR	Operational Concept Document (OCD)	Reviewed	
SDR	Preliminary Development Specification (Type B)	Reviewed	
SDR	Preliminary Software Requirements Specification (SRS) (Type B5)	Reviewed	
SDR	Preliminary Interface Requirements Specification (IRS) (Type B5)	Reviewed	
SDR	System/Segment Design Document (SSDD)	Reviewed	
SDR	Preliminary CI-Level Interface Control Documents (ICD)	Reviewed	
SDR	Preliminary Requirements Matrix	Reviewed	
SDR	Preliminary Technical Manuals	Reviewed	
SDR	Drawings	Reviewed	
DM	Standards and Specifications	Appvd for Use by SEM	

Table 7-1B   *Requirements Analysis Output Checklist*

Customer	Output	Exit Criteria	Task #	Status
SE Mgmt	Engineering Change Proposals (ECP)	Inspected and Appvd by CAM	Any	
SE Mgmt	Modification Request (MR), Change Request (CR)	Appvd by CAM	Any	
ALL	System Scenarios	Inspected	303	
Fn Anal/Alloc, SI&V	Mission Profile	Inspected	302, 301	
Sys Anal & Opt	Operational Profile	Inspected	302, 301	
ALL	House of Quality Diagram	Inspected	301	
ALL	Design-To-Cost (DTC) Goals	Appvd by SEM	305	
ALL	Cost Breakdown Structure (CBS)	Appvd by SEM	305	
Fn Anal/Alloc	Environmental Constraints	Inspected	305	
Fn Anal/Alloc	Design Constraints	Inspected	305	
Fn Anal/Alloc, SI&V	Operational Concept Document (OCD)	Inspected	306	
Fn Anal/Alloc, SI&V	Measures of Effectiveness (MOEs)	Inspected	307	
ALL	Simulation Objectives	Appvd by SEM	307, 308	
Fn Anal/Alloc	Performance Requirements	Inspected	308	
Fn Anal/Alloc, SI&V	Technical Parameters (TPs)	Inspected	308	
Sys Anal & Opt	Effectiveness Analysis Request	Appvd by SEM	Any	
Sys Anal & Opt	Trade Study Request	Appvd by SEM	Any	

7.2 *Functional Analysis/Allocation Activity*

"Gold is where you find it."

—American saying

System elements should not be defined or acquired without first justifying their need through the functional analysis activity. The tasks for this activity are shown in Figure 7-3.

The system objectives defined in Requirements Analysis are analyzed to define the required functional behavior of the system. This analysis may be through Functional Flow Block Diagrams (FFBD), Timelines, Data Flow Diagrams, State/Mode Diagrams, Behavior Diagrams or other diagramming techniques. The allocation part of the task establishes traceability between requirements, functions and system elements. Requirements are allocated to functions and budgeted among them. The functional allocations may be documented on Requirements Allocation Sheets (RAS).

The functional block diagram approach should include coverage of all activities throughout the system life cycle, and the method of presentation should reflect proper activity sequences and interface relationships.

The information included within the functional blocks should be concerned with "what" is required before looking at "how" it should be accomplished.

The diagramming technique should be flexible enough to allow for expansion when additional definition causes too much detail to be presented at once. The objective is to progressively and systematically work down to the level where resources can be identified with "how" a task or function should be accomplished.

Functions will be developed concurrently and iteratively to define functional relationships and dependencies, including internal and external functional interfaces. Outputs from other SE process tasks will be used to support further decomposition of functions, and in defining alternative subfunctions to satisfy higher level functions.

Automated simulation tools should be used wherever possible in accomplishing functional analysis and allocations. Logistics Support Analysis (LSA) will be performed throughout this activity. Process requirements for manufacturing, verification and support may be identified through factory and support simulations.

All requirements, including those derived from simulation, should be documented as to source.

Task 401 Define System States & Modes

Through analysis of the system's expected environment and intended uses, define the states and modes that the system will experience. The states and modes should be consistent with the mission profile and the concept of operations. For example, changes of physical states may occur due to temperature changes during the deployment operations. Life cycle activities, including production and deployment, should be defined with flow charts and exit criteria at the system level and at successively lower levels in the architecture with each iteration through the Requirements and Architecture Definition Subprocess.

Task 402 Define System Functions

Derive the operational and support behavior of the system in terms of the functions the system must perform. This task includes some form of functional flow, control flow or data flow analyses. Sometimes it is necessary to define the behavior of external systems in order to fully characterize the external interfaces.

    The functions and the functional interfaces may be documented on a Functional Flow Block Diagram (FFBD) function lists, or function trees. Describe each function using a "verb" to define the required task, action or activity.

Task 403 Define Functional Interfaces

Define the start/end states and inputs/outputs for each function. This ensures that all state transitions are completely defined within functions, and required inputs and outputs are provided. Identify the interface items that trigger each function. Review system functionality for completeness and consistency before performance is allocated.

    Interaction of the system with other systems in the customer premises or organizational structure may require special interfaces. Existing transportation and handling equipment may dictate interfaces such as tie-downs and design constraints such as maximum weight, volume and size.

Task 404 Define Performance Requirements and Allocate to Functions

For each function defined, determine how well that function must be performed. The required system performance should drive the individual function performance. Identify the technical budgets that need to be established. Model performance allocations as appropriate to assist in the allocation process. Records of the decisions and trade studies generated during this task may be recorded on Requirements Allocation Sheets (RAS) to ensure traceability is maintained.

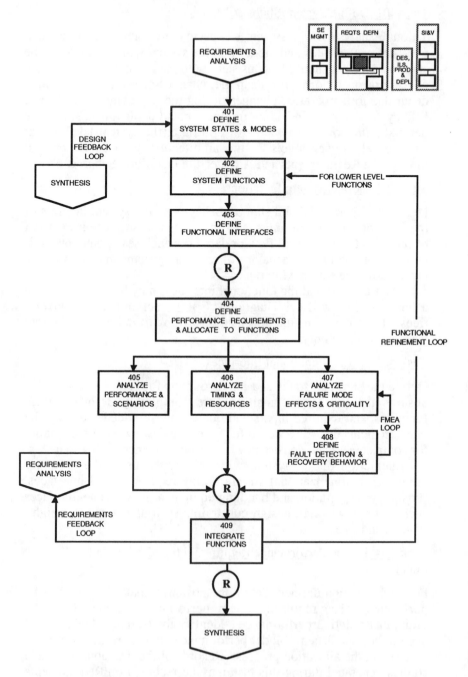

Figure 7-3 *Functional Analysis/Allocation Activity*

Task 405 Analyze Performance and Scenarios

Using the functional description, analyze the modeled behavior for static and dynamic consistency and executability. Use modeling and simulation tools to the maximum extent possible to help assess expected performance of individual system elements and the system as a whole. These models are developed in Task 601.

For each system scenario analyze the system behavior. For each system stimulus analyze the system response. Define execution "threads" through the functional architecture. These threads can be used during verification to establish that proper functions are being performed.

Task 406 Analyze Timing & Resources

Using the functional description and system constraints, analyze the system behavior for compliance with timing requirements and internal executability from a time and resource utilization aspect.

Task 407 Analyze Failure Mode Effects and Criticality

Analyze the functional consequences of any specific failure. Do this in conjunction with the Failure Mode Effects and Criticality Analysis (FMECA) performed by reliability and maintainability engineers as appropriate. Specific operational or mission profile information is usually required to support the FMECA. The results of this task may be documented in fault trees or FMEA tables.

Task 408 Define Fault Detection & Recovery Behavior

Provide modifications to the functional definition to respond to out-of-normal conditions. Fault recovery from operational failures may lead to the need for maintenance functions. Analyze the new behavior again for failure mode effects and criticality.

Task 409 Integrate Functions

Assure that all functions collectively provide optimum system performance according to the defined Measures of Effectiveness (MOEs) and meet system requirements. Repeat the above tasks for lower level functions.

Assure that system and subsystem interfaces are defined for all interfaces: physical, functional, fit, data format, support, etc.

Table 7-2A   Functional Analysis/Allocation Input Checklist

Supplier	Input	Entry Criteria	Status
Reqts Anal	Mission Profile	Inspected	
Reqts Anal	Environmental Constraints	Inspected	
Reqts Anal	Design Constraints	Inspected	
Reqts Anal	Measures of Effectiveness (MOEs)	Inspected	
Reqts Anal	Technical Parameters (TPs)	Inspected	
Reqts Anal	Operational Concept Document (OCD)	Inspected	
Reqts Anal	Performance Requirements	Inspected	

Table 7-2B   Functional Analysis/Allocation Output Checklist

Customer	Output	Exit Criteria	Task #	Status
SE Mgmt	Engineering Change Proposals (ECP)	Inspected and Approved by CAM	Any	
SE Mgmt	Modification Request (MR), Change Request (CR)	Approved by CAM	Any	
Synthesis	State/Mode Diagrams	Inspected	401	
Synthesis, Dev Teams	Life Cycle Definition Report	Inspected	401	
Synthesis	Functional Flow Block Diagrams (FFBD)	Inspected	402	
Synthesis	Function Trees	Inspected	402	
Synthesis	Data Flow Diagrams	Inspected	402	
Synthesis	Control Flow Diagrams	Inspected	402	
Synthesis	Performance Requirements	Inspected	404	
Synthesis	Requirements Allocation Sheets (RAS)	Inspected	404	
Synthesis	Technical Budgets	Inspected	404	
Synthesis	Time Line Sheets (TLS)	Inspected	406	
Synthesis	Timing/Resource Analysis Report	Inspected	406	
Synthesis	Failure Mode and Effects Analysis (FMEA) Tables	Inspected	407	
Synthesis	Failure Trees	Inspected	407	
Synthesis, SI&V	Test Requirements Sheets (TRS)	Inspected	Any	
Synthesis, ILS	End Item Maintenance Sheets (EIMS)	Inspected	Any	

Synthesis, ILS	Logistic Support Analysis Records (LSAR)	Inspected	Any	
Synthesis, Prod	Production Sheets (PS)	Inspected	Any	
Sys Anal & Opt	Effectiveness Analysis Request	Appvd by SEM	Any	
Sys Anal & Opt	Trade Study Request	Appvd by SEM	Any	

7.3 Synthesis Activity

"Man is made to create, from the poet to the potter."
                                        —Benjamin Disraeli

The Synthesis Activity (Figure 7-4) defines the product architecture that meets the functional and performance requirements. At the lowest level, synthesis defines the "build-to" requirements for the system elements which are designed by the development teams. The Configuration Items (CIs) are established and defined during this activity. At each level in the architecture, design requirements, process requirements, physical configuration and interfaces must be verified to ensure that functional requirements are satisfied.

Synthesis is conducted to define system elements and to refine and integrate them into a physical configuration of the system that satisfies functional requirements.

Alternate configurations, or architectures, are developed and evaluated against the requirements. Prototypes or models may be constructed for more than one architecture to support trade-off analysis of valid alternatives.

This activity is related to the "systems architecting" activity described in [Rechtin, 1991] and [Rechtin, 1996]. Many of the detailed engineering methods related to this activity are described in [Blanchard, 1990].

Task 501 Assess Technology Alternatives

Evaluate technologies that can be applied to the problem. Include all possibilities referred to in the Technology Insertion Plan. Identify possible system concepts and options. Examine technology trends to determine level of technology appropriate at time of deployment.

Task 502 Synthesize System Element Alternatives

Define and refine system element alternatives for each logical set of functional requirements using a bottom-up approach.

Task 503 Allocate Functions to System Elements

Identify which functions will be performed by which system elements. Allocate the associated performance of each function to the appropriate system element.

Task 504 Allocate Constraints to System Elements

Identify constraints that apply directly to system elements, and that do not apply to behavioral functions. These constraints include such "non-functional" requirements as power, volume, weight, dimensions, etc.

Task 505 Define Physical Interfaces

Define mechanical, electrical, data, and other interfaces between elements of the system. Identify all interfaces between the system and the outside world. Document these interfaces in interface control documents or other interface agreements.

Depict all elements of the system, and interactions among them and the customer environment in component diagram form. Include existing logistics support equipment. A typical format used here is the Schematic Block Diagram (SBD).

Task 506 Define Platform and Architecture

Define the platform(s) upon which the product will reside. This could be an item already existing in the user premises or to be provided along with the delivered product. The platform could be common across a family of products.

Define each architecture in terms of product structures and interactions between the products and with things in the environment. Map scenarios (Task 303) to various configurations of the system.

The hierarchical relationship of the system elements should be documented in an Architecture Block Diagram (ABD). The ABD should include the hardware and software "family trees," documentation and data, facilities, special test equipment, etc. See [Grady, 1993] for examples of an ABD.

An external ABD should be developed which includes the external elements which affect the system under development. (These external items are the enabling products as described in Chapter 2— development, production, deployment, test, support, training, and disposition.) The external ABD should include all hardware, software, facilities, personnel, data, and services which may have a significant effect on your system.

Define the specific building block elements for the product. (See Section 2.2.1 for a description of the generic building block to be used in engineering of systems.)

Task 507 Refine Work Breakdown Structure (WBS)

Translate the selected architecture (per the ABD and the basic building block) and its decomposition into WBS format for work planning, and cost/schedule tracking and control. This task will extend the existing WBS to account for the additional system elements synthesized.

Task 508 Develop Life Cycle Techniques & Procedures

Develop appropriate model and parameters for life cycle cost analysis. Define how the system will be manufactured, verified, deployed, operated, maintained and disposed of. Define training and other ILS procedures.

Identify all required enabling products and key characteristics of these enabling products. Define the requirements imposed on the enabling products by the end products. Identify constraints of these enabling products that will be imposed on the end products.

Task 509 Check Requirements Compliance

Ensure that all functional and performance requirements have been mapped to the system elements. Ensure that the system elements at each level in the architecture satisfy all requirements and constraints. Compliance may be checked using System/Cost Effectiveness Analyses, simulations, demonstrations, inspections, or tests. The models and prototypes produced in Task 601 may also be used.

Task 510 Integrate System Elements

Progressively integrate (bottom-up) the system elements into items that provide an end-use function. At each level, the resulting design requirements, physical configuration and physical interfaces should be verified to ensure that functional and performance requirements are satisfied.

The integration activity described here is "logical" in the sense that the elements are not physically integrated since this occurs during the SI&V activities. The integration could be done using analysis tools and techniques, CAD tools, simulation tools, etc.

Task 511 Select Preferred Design

From the system element alternatives, select the system element— using trade study results—that leads to a balanced system solution.[37] Coordinate the design selection with the SI&V Plan.

---

[37]It is best not to optimize the subsystems individually. This almost always leads to a less than optimal system.

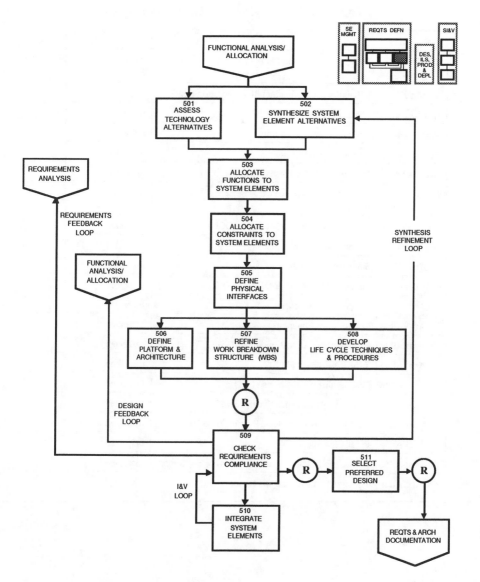

Figure 7-4   *Synthesis Activity*

Table 7-3A   Synthesis Input Checklist

Supplier	Input	Entry Criteria	Status
SE Mgmt	Technology Insertion Plan	Approved by SEM	
SE Mgmt	Reliability and Maintainability Allocation Report	Inspected	
Reqts Anal	Mission Profile	Inspected	
Reqts Anal	Environmental Constraints	Inspected	
Reqts Anal	Design Constraints	Inspected	
Reqts Anal	Measures of Effectiveness (MOEs)	Inspected	
Reqts Anal	Technical Performance Measurement (TPM) Parameters	Inspected	
Reqts Anal	Operational Concept Document (OCD)	Inspected	
Reqts Anal	Performance Requirements	Inspected	
Fn Anal/Alloc	State/Mode Diagrams	Inspected	
Fn Anal/Alloc	Functional Flow Block Diagrams (FFBD)	Inspected	
Fn Anal/Alloc	Data Flow Diagrams	Inspected	
Fn Anal/Alloc	Control Flow Diagrams	Inspected	
Fn Anal/Alloc	Requirements Allocation Sheets (RAS)	Inspected	
Fn Anal/Alloc	Time Line Sheets (TLS)	Inspected	
Fn Anal/Alloc	Test Requirements Sheets (TRS)	Inspected	
Fn Anal/Alloc	End Item Maintenance Sheets (EIMS)	Inspected	
Fn Anal/Alloc	Logistic Support Analysis Records (LSAR)	Inspected	
Fn Anal/Alloc	Production Sheets (PS)	Inspected	
Fn Anal/Alloc	Failure Mode and Effects Analysis (FMEA) Report	Inspected	
Fn Anal/Alloc	Timing/Resource Analysis Report	Inspected	

Table 7-3B   Synthesis Output Checklist

Customer	Output	Exit Criteria	Task #	Status
SE Mgmt	Engineering Change Proposals (ECP)	Inspected and Appvd by CAM	Any	
SE Mgmt	Modification Request (MR), Change Request (CR)	Appvd by CAM	Any	
Sys Elem Doct	Concept Description Sheets (CDS)	Inspected	Any	
Sys Elem Doct	Design Description Sheets	Inspected	502	
Sys Elem Doct	Requirements Allocation Sheets (RAS) with Component Allocations	Inspected	503, 504	
Sys Elem Doct	Design Constraint Sheets (DCS)	Inspected	504	
Sys Elem Doct	Schematic Block Diagrams (SBD)	Inspected	505	
Sys Elem Doct	Interface Drawings	Inspected	505	
Sys Elem Doct	Interface Worksheets	Inspected	505	
Sys Elem Doct	Facility Interface Sheets	Inspected	505	
Sys Elem Doct, SE Mgmt	Architecture Block Diagram (ABD)	Inspected	506	
Sys Elem Doct	Specification Tree	Inspected	506	
SE Ctl & Integ	Extended/Updated WBS	Inspected	507	
SE Ctl & Integ	Requirements Compliance Matrix	Inspected	509	
Sys Anal & Opt	Effectiveness Analysis Request	Appvd by SEM	Any	
Sys Anal & Opt	Trade Study Request	Appvd by SEM	Any	

7.4 *System Analysis and Optimization Activity*

"If there be light, then there is darkness; if cold, then
heat; if height, depth also; if solid, then fluid; hardness
and softness; roughness and smoothness; calm and
tempest; prosperity and adversity; life and death."
                                        —Pythagoras

The System Analysis and Optimization Activity (Figure 7-5) provides
the risk assessment, progress measurement, and decision mechanism
required to evaluate design capabilities, determine progress in
satisfying technical requirements and Design-To-Cost (DTC)
objectives, formulate and evaluate alternative courses of action, and
evolve the total system to satisfy customer's requirements.

Effectiveness analyses include Production Engineering Analysis,
Test and Verification Analysis, Deployment & Installation Analysis,
Operational Analysis, Supportability Analysis, Training Analysis,
Disposal Analysis, Environmental Analysis, and Life Cycle Cost
(LCC) Analysis.

Performance-based progress measurement and criteria are
documented in the Systems Engineering Master Schedule (SEMS), and
include Technical Performance Measurement (TPM) and technical
reviews and audits.

Task 601 Develop System Models

Models are developed in order to characterize the expected behavior of
a system to ensure that it will adequately meet the users' needs and
expectations. The models could be logical or physical. Models are
used to discover unexpected behavior in order to shorten the
development time and to mitigate unknown or known development
risks.

Define the questions to be addressed by each model. If the models
are to be used in demonstrations to potential customers or users, define
the objectives of the demonstrations. Define acceptable behavior for
the model. Define degree of fidelity required in the model or in the
results of test or analysis using the model. Define intended users and
uses for the model.

Develop models for specific characteristics of system
functionality. These include, but are not limited to: operations (such
as measures of effectiveness), supportability, reliability,
maintainability, production, training, disposal, test/verifications,
deployment/installation, environmental, and life cycle costs (including

DTC). Validate the system model to ensure valid analysis and simulation results.

These models may include physical prototypes, mathematical formulations, simulation models, breadboards, brassboards, mass mockups, scale models, wire-frame or solids models in CAD tools, executable code frames, etc.

Task 602 Perform System Effectiveness and Cost Effectiveness Analysis

Using validated models, analyze the proposed solution or derived requirements. Determine sensitivity to constraints. These analyses assist in the identification of parameters that drive solutions and establish their sensitivity to uncertainties in input data and assumptions. These effectiveness analyses are performed to:

a) support the identification of mission and performance objectives and requirements

b) support the allocation of performance to functions

c) provide criteria for the selection of solution alternatives

d) provide analytic confirmation that designs satisfy customer requirements

e) support verifications of people, product and process solutions.

Task 603 Risk Evaluation

Perform a risk analysis on the proposed requirement, function, or solution for use in the trade study or updating the problem risk analysis. Include producibility and supportability requirements. Assess risk to planned profits based on the evolving requirements versus contractually covered requirements.

Task 604 Trade Studies

Decision criteria will be established and the alternatives evaluated against criteria. The criteria must include any customer provided MOE or Cost and Operational Effectiveness Analysis (COEA) alternatives. A weighted scoring method may be applied including a sensitivity analysis. However, the final decision is not restricted to the best score and may include other factors. Use decision trees to develop a trade study hierarchy and to assist in simplifying the trade space.

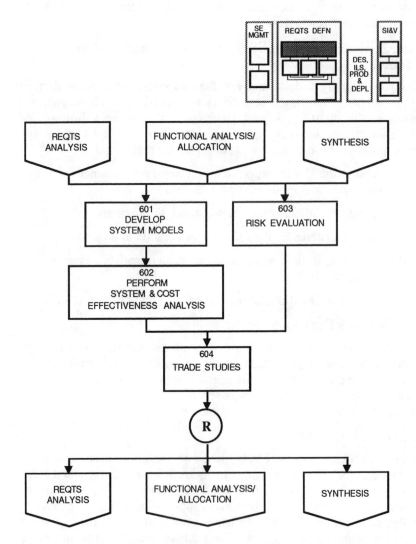

Figure 7-5 *System Analysis and Optimization Activity*

Table 7-4A **System Analysis and Optimization Input Checklist**

Supplier	Input	Entry Criteria	Status
Several	Effectiveness Analysis Request	Approved by SEM	
Several	Trade Study Request	Approved by SEM	
Several	All Outputs from Requirements Analysis, Functional Analysis/Allocation and Synthesis	Various	
SE Mgmt	Systems Engineering Development Environment (SEDE) Plan	Approved by SEM	
SE Mgmt	Tailored SE Process Document	Approved by SEM	
SE Mgmt	Effectiveness and Trade Studies Plan	Approved by SEM	
SE Mgmt	Technology Insertion Plan	Approved by SEM	
SE Mgmt	Risk Assessment Report	Approved by SEM	
SE Mgmt	TPM Charts	Approved by SEM	
SE Mgmt	Specialty Engineering Plans	Approved by SEM	
SE Mgmt	ILS Plan	Approved by SEM	
SE Mgmt	Producibility Plan	Approved by SEM	
SE Mgmt	Deployability Plan	Approved by SEM	
SE Mgmt	Risk Management Plan	Approved by SEM	
SE Mgmt	Other Plans	Approved by SEM	

Table 7-4B **System Analysis and Optimization Output Checklist**

Customer	Output	Exit Criteria	Task #	Status
SE Mgmt	Engineering Change Proposals (ECP)	Inspected and Approved by CAM	Any	
SE Mgmt	Modification Request (MR), Change Request (CR)	Approved by CAM	Any	
Development Teams, SI&V	System Model	Validated	601	
SI&V, SE Mgmt	System Model Validation Report	Reviewed	601	
Requester, SE Mgmt	Effectiveness Analysis Report	Reviewed	602	
Development Teams, SE Mgmt	Risk Evaluation Report	Reviewed	603	
Requester, SE Mgmt	Trade Study Report	Reviewed	604	

7.5 Requirements and Architecture Documentation Activity

The resulting set of design requirements from the Synthesis Activity will be the basic source of data for developing, establishing and updating the functional, allocated, and product baselines and the documents listed below. These documents are defined in the Glossary. Figure 4-5 shows the relative timing of the different specifications. The tasks for this activity are shown in Figure 7-6.

Requirements and Architecture Documentation may be developed in accordance with MIL-STD-961C, Specification Practices, which defines different types of specifications.

- A Spec, System/Segment Specification

- B Spec, Development Specification

- C Spec, Product Function/Fabrication Specification

- D Spec, Process Specification

- E Spec, Material Specification

- System/Segment Design Document (SSDD)

J-Std-016 defines software specification types including the following:

- Software Requirements Specification (SRS)

- Interface Requirements Specification (IRS)

- Software Design Document (SDD)

- Interface Design Document (IDD)

- Software Product Specification (SPS)

Other types of documentation developed are the following:

- Interface Control Document (ICD)

- Test Requirements Specification (TRS)

- Requirements Traceability Matrix (RTM)

- Requirements Verification Matrix (RVM)

- Drawings

- Technical manuals[38]

- Training material

Task 701 Develop Document Approach

Identify customer needs for this document. Use Data Item Descriptions (DIDs) and organizational standard templates for guidance. Collect examples of similar documents to be used for guidance.

Task 702 Develop Detailed Document Outline

Adapt standard format to the needs of the project. Inspect the detailed outline to ensure compliance with customer requirements, DIDs, and with organizational standard templates. Identify all inputs required for the document (see Table 7-5A).

Task 703 Develop Text

Develop the textual content for the document draft. Review the draft text with document customers, internal and external. Update per customer comments.

Task 704 Develop Graphics

Create figures for the document from material produced during the Requirements and Architecture Definition Subprocess (see Table 7-5A). Review the draft graphics with document customers, internal and external. Update per customer comments.

Task 705 Produce Document

Integrate document elements and perform technical editing. Include review comments and produce document. Perform peer inspection in accordance with In-Process Quality Inspection (IPQI) process and incorporate peer inspection comments in document.

Task 706 Deliver Document

After sign-off by SE management and appropriate managers, enter into CM database and execute document release process. Data Management (DM) will transmit to customer per contract requirements with a cover letter signed by the PM. DM will deliver to internal customers and to project library.

---

[38]Sometimes technical manuals and training material can be used to document the requirements and architecture of the product. This approach can sometimes be used in lieu of the conventional specification document approach.

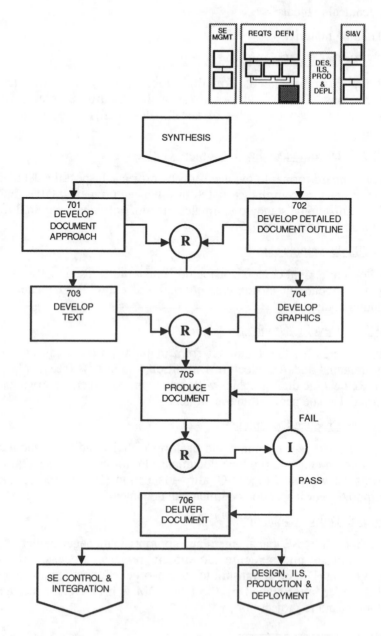

Figure 7-6 *Requirements and Architecture Documentation Activity*

Table 7-5A   **Requirements and Architecture Documentation Input Checklist**

Supplier	Input	Entry Criteria	Status
Data Mgmt	Standards and Specifications	Approved for Use by SEM	
Data Mgmt	Data Item Descriptions (DIDs)	Approved for Use by SEM	
SE Mgmt	Approved Engineering Change Proposals (ECP)	Approved by Customer CCB	
SE Mgmt	Approved Change Orders (CO)	Approved by Project CCB	
Reqts Anal	Requirements Traceability Matrix (RTM)	Inspected	
Reqts Anal	Mission Profile	Inspected	
Reqts Anal	Operational Concept Document (OCD)	Inspected	
Fn Anal/Alloc	Functional Flow Block Diagrams (FFBD)	Inspected	
Fn Anal/Alloc	Data Flow Diagrams	Inspected	
Fn Anal/Alloc	Control Flow Diagrams	Inspected	
Fn Anal/Alloc	Time Line Sheets (TLS)	Inspected	
Fn Anal/Alloc	Test Requirements Sheets (TRS)	Inspected	
Synthesis	Concept Description Sheets (CDS)	Inspected	
Synthesis	Requirements Allocation Sheets (RAS)	Inspected	
Synthesis	Design Constraint Sheets (DCS)	Inspected	
Synthesis	Schematic Block Diagrams (SBD)	Inspected	
Synthesis	Architecture Block Diagram (ABD)	Inspected	
Synthesis	Interface Drawings	Inspected	
Synthesis	Interface Worksheets	Inspected	

Table 7-5B   *Requirements and Architecture Documentation Output Checklist*

Customer	Output	Exit Criteria	Task #	Status
ALL	System/Segment Specification (Type A)	Note 1	706	
ALL	Development Specification (Type B)	Note 1	706	
ALL	Product Specification (Type C)	Note 1	706	
ALL	Process Specification (Type D)	Note 1	706	
ALL	Material Specification (Type E)	Note 1	706	
ALL	System/Segment Design Document (SSDD)	Note 1	706	
ALL	Software Requirements Specification (SRS) (Type B5)	Note 1	706	
ALL	Interface Requirements Specification (IRS) (Type B5)	Note 1	706	
ALL	Software Design Document (SDD) (Type C5)	Note 1	706	
ALL	Interface Design Document (IDD) (Type C5)	Note 1	706	
ALL	Software Product Specification (SPS) (Type C5)	Note 1	706	
ALL	Interface Control Document (ICD)	Note 1	706	
ALL	Test Requirements Specification (TRS)	Note 1	706	
ALL	Requirements Traceability Matrix (RTM)	Note 1	706	
ALL	Requirements Verification Matrix (RVM)	Note 1	706	
ALL	Technical Manuals	Note 1	706	
ALL	Training Material	Note 1	706	
Production, Deployment	Drawings	Note 1	706	

NOTE 1–All documents must be inspected, approved by the Systems Engineering Manager and QA, and under CM Control.

References

Blanchard, B., and W. Fabrycky, *Systems Engineering and Analysis*. Prentice Hall, 1990.

Grady, Jeffrey O., *System Requirements Analysis*. McGraw Hill, 1993.

Rechtin, Eberhardt, *Systems Architecting: Creating and Building Complex Systems*. Prentice Hall, 1991.

Rechtin, Eberhardt and Mark Maier, *The Art of Systems Architecting*. CRC Press, 1996.

chapter eight

System integration and verification (SI&V) subprocess

"There is always an easy solution to every human problem—neat, plausible, and wrong."

—H. L. Mencken

The System Integration and Verification Subprocess (Figure 8-1) in its broadest meaning describes all activities performed by developers and testers that pertain to integrating and verifying a product from the smallest unit through the complete system. Whatever development methodology applies to a project, the development should proceed in accordance with an overall master plan that defines how, when, and by whom the product will be synthesized into its final deliverable form and verified to meet its requirements.

As used within the SE process, SI&V pertains to the full scope of integration and verification activities that occur as the product(s) being developed under the project near completion. The products, as defined in the SOW, whether hardware, software, or processes, must be integrated and verified against their requirements in order to ensure the customer's acceptance of them. The Systems Engineering Management Plan (SEMP) defines at a high level how the products and processes will be developed and integrated. The Systems Engineering Detailed Schedule (SEDS) defines when the products, services, and processes will be integrated and verified.

The Software Development Plan (SDP) is often developed by the software development team and provided to the SI&V subprocess. The SDP will address how software is to be integrated with the hardware.

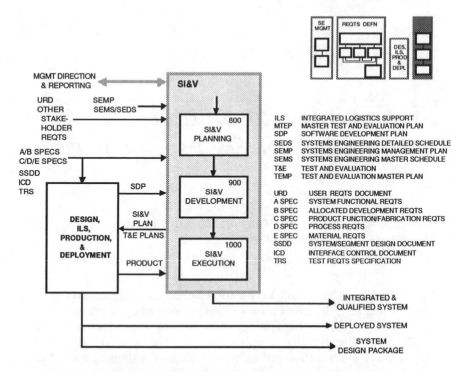

Figure 8-1   *System Integration and Verification (SI&V) Subprocess*

A Test and Evaluation Master Plan (TEMP) may be provided by the customer to define the overall test and evaluation plan the customer will use to evaluate the contractor's product or service. If fully implemented on a project, SI&V covers all of the activities depicted in Figure F-1 in boxes 60 through 66 (see Appendix F). However, many projects are likely to require only a subset of these activities and would tailor SI&V appropriately.

All SI&V activities are driven by the requirements which are generated in the SE process. Requirements are mapped via a Requirements Traceability Matrix (RTM) through the hierarchy of specifications to the specific system elements that apply. The Requirements Verification Matrix (RVM) [39] maps all requirements into specific verification methods. The methods of analysis, inspection, demonstration and test are the focus of the verification activity.

---

[39]The Requirements Verification Matrix (RVM) is usually included in Section 4 of a specification.

The Master Test and Evaluation Plan (MTEP) identifies the full scope of the test program, mapping the requirements into particular test categories—for example, system acceptance testing, reliability growth testing, environmental testing, preproduction qualification testing, etc.

8.1 SI&V Planning Activity

The SI&V Planning Activity (Figure 8-2) consists of the formulation of multiple plans. These plans define the overall SI&V program to be conducted, including all elements of integration and verification to be performed from development level tests through final product acceptance. In addition, each category of verification, including System Qualification, Preproduction Qualification, Product Acceptance, Predeployment Qualification, and Site Acceptance, typically requires its own plan that focuses on the unique requirements ascribed to the category. The overall SI&V program is usually defined in a System Integration and Verification (SI&V) Plan.

Each plan identifies the environments needed to integrate and verify the various subsystems into the final product or service, establishes the verification approach that will be followed, and defines the high level procedures that will be performed. Because of the criticality of SI&V to the customer, these plans often require customer approval before the SI&V program may proceed.

Task 801 Analyze Customer and Higher Level I&V Plans

Analyze customer requirements, project-related plans, and documentation including the Technical Proposal and subcontractor training agreements. Test requirements may be in the form of a Test and Evaluation Master Plan (TEMP) or simply included as part of the Statement of Work (SOW). From these documents the critical test and evaluation issues should be identified. The critical T&E issues will drive the entire development.

If subcontractors are to be involved in the project, the project plan should be provided for development of their plans. The subcontractor plans must be reviewed for their consistency with overall project guidelines and system/equipment requirements. All project-specific documentation should be gathered, reviewed, and used as input to prepare or update the SI&V plan.

Task 802 Develop RVM

Develop a Requirements Verification Matrix (RVM) as a report from the requirements database using the RTM provided by the Requirements and Architecture Definition Subprocess. The RVM, for each requirement, assigns a verification method, i.e., test, demonstration, analysis, or inspection, and test case reference (if applicable), and provides a compliance column that will reference the documented results after the requirement is verified.

Prepare a RVM for Section 4 of each specification. There could be one RVM for the entire system including its elements, or there could be one RVM for each system element.

A Requirements Verification Traceability Matrix (RVTM) is often developed which is merely a combination of both the RTM and RVM into one matrix.

Task 803 Define Integration Requirements

Define integration requirements to validate the final integrated product. These may take the form of a checklist of predecessor relationships. Steps in the checklist may reference other categories of testing such as System Qualification, Preproduction Qualification, Product Acceptance, Predeployment Qualification, and Site Acceptance or reference assembly drawings and thus constitute the integration procedures.

Task 804 Develop I&V Strategy/Approach

Develop an I&V strategy/approach to address issues such as test versus analysis, black-box (input-output driven) versus white-box (logic driven) testing, exhaustive versus minimum/nominal/maximum testing, and synthetic versus real-world data. Also develop requirements for facility needs, staffing/discipline needs, data needs, and equipment (hardware and software) needs.

Task 805 Define Verification Requirements

Define verification requirements to validate the final integrated product. These may take the form of a checklist of predecessor relationships. Steps in the checklist may reference other categories of testing such as System Qualification, Preproduction Qualification, Product Acceptance, Predeployment Qualification, and Site Acceptance or reference assembly drawings and thus constitute the verification procedures.

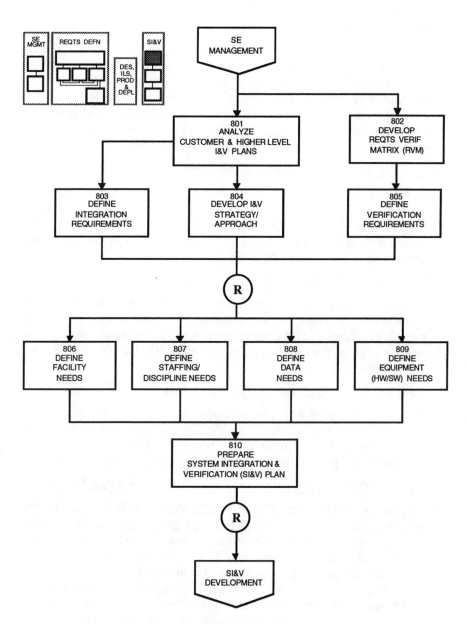

Figure 8-2 *SI&V Planning Activity*

Task 806 Define Facility Needs

Define facility needs for the software or hardware development environment, the laboratory test environment, and the configuration control environment. Software development may require a host computer(s) with a user interfacing network, while hardware development may require a design control computer. Software and hardware development require laboratory and configuration control environments for testing and controlling the product.

Task 807 Define Staffing/Discipline Needs

Define staffing/discipline needs for product design, development, testing, and delivery to the customer. These include System Engineering Team, Software/Hardware Development Team, Test Team, Software Product Administrator, Facility Administrator, and Quality Assurance Engineer(s).

Task 808 Define Data Needs

Define data needs such as simulation/stimulation data, drivers, and test applications for product testing which is the mechanism for acceptance by the customer. These data needs enable the functionality and performance of the product to be determined in the absence of real-world input data.

Task 809 Define Equipment (HW/SW) Needs

Define the equipment (hardware and software) needs for integration and test. These needs encompass specification of the target hardware and/or application software under test and ancillary hardware and software to support multiple layers of integration and verification.

Task 810 Prepare System Integration and Verification (SI&V) Plan

Prepare the SI&V Plan. The purpose of the SI&V Plan is to identify all developmental subsystem and system integration and verification activities performed by the prime contractor and subcontractors, identify all plans to be developed, by discipline, and define their interrelationships, identify requirements allocation to each of the plans, identify the SI&V program schedule, showing the time phasing of all SI&V activities, identify required resources, identify risk areas and contingency resources, and identify customer participation requirements for testing.

Table 8-1A SI&V Planning Input Checklist

Supplier	Input	Entry Criteria	Status
SE Mgmt	Systems Engineering Management Plan (SEMP)	Appvd by SEM	
SE Mgmt	Systems Engineering Master Schedule (SEMS)	Appvd by SEM	
SE Mgmt	Systems Engineering Detailed Schedule (SEDS)	Appvd by SEM	
SE Mgmt	User Requirements Document (URD)	Appvd by SEM	
SE Mgmt	Test and Evaluation Master Plan (TEMP)	Appvd by SEM	
SE Mgmt	Tailored SE Process Document	Appvd by SEM	
SE Mgmt	Effectiveness and Trade Studies Plan	Appvd by SEM	
SE Mgmt	TPM Charts	Appvd by SEM	
SE Mgmt	Technical Review Plan	Appvd by SEM	
SE Mgmt	Technology Insertion Plan	Appvd by SEM	
SE Mgmt	Configuration Management Plan	Appvd by SEM	
SE Mgmt	Specialty Engineering Plans	Appvd by SEM	
SE Mgmt	ILS Plan	Appvd by SEM	
SE Mgmt	Producibility Plan	Appvd by SEM	
SE Mgmt	Deployability Plan	Appvd by SEM	
SE Mgmt	Risk Management Plan	Appvd by SEM	
SE Mgmt	Risk Assessment Report	Appvd by SEM	
SE Mgmt	Other Plans	Appvd by SEM	
SE Reqts Defn	A/B/C/D/E Specifications	Appvd by SEM	
SE Reqts Defn	Interface Control Documents (ICDs)	Appvd by SEM	
SE Reqts Defn	System/Segment Design Document (SSDD)	Appvd by SEM	
SE Reqts Defn	Test Requirements Specification (TRS)	Appvd by SEM	
SE Reqts Defn	Technical Parameter (TP) Metrics	Inspected	
SE Reqts Defn	Technical Performance Measurement (TPM) Parameters	Appvd by SEM	
SE Reqts Defn	Operational Concept Document (OCD)	Appvd by SEM	
SE Reqts Defn	Mission Profile	Appvd by SEM	
SE Reqts Defn	Measures of Effectiveness (MOEs)	Appvd by SEM	
SE Reqts Defn	Requirements Traceability Matrix (RTM)	Inspected	
SW Dev Team	Software Development Plan (SDP)	Appvd by SEM	
Deployment	Deployment Procedures	Inspected	
ILS	Technical Manuals	Inspected	

Table 8-1B   SI&V Planning Output Checklist

Customer	Output	Exit Criteria	Task #	Status
ALL	Requirements Verification Matrix (RVM)	Inspected	802	
PM	SI&V Facility Request	Appvd by SEM	806	
PM	SI&V Staffing Request	Appvd by SEM	807	
PM	SI&V Data Request	Appvd by SEM	808	
PM	SI&V Equipment Request	Appvd by SEM	809	
ALL	System Integration and Verification (SI&V) Plan	Appvd by SEM	810	

8.2 SI&V Development Activity

SI&V Development consists of four sets of tasks (see Figure 8-3):

1. Defining the specific methods that will verify the functional and performance requirements allocated to the various subsystems to be integrated or verified, where each method identifies the function (feature) to be tested, the requirement(s) to be verified, the general procedure that will be followed, and the pass criteria for the test. The Requirements Verification Matrix (RVM) will be updated to reflect the mapping of requirements to test cases. The hand-off criteria for the acceptance of each subsystem from the design organization are defined and inspected with the designers to ensure their understanding of the required level of integrity expected of the product at the time it is delivered for SI.

2. Defining and establishing the environment within which the integration or verification activity will be performed. This includes, for example, the facility, any special drivers or tools, any special software or hardware needed to achieve the desired test conditions, and any data sources needed to drive the system.

3. Defining the detailed, step-by-step procedures for each verification activity. Procedures such as Taguchi design of experiments may be employed to minimize unnecessary test events. Typical test procedures identify for each test the actual sequence of commands and the responses to them expected from the product under test. When formal tests are executed for the customer, the procedures serve as a checklist to ensure that the test is performed correctly.

4. Documenting the above information and obtaining customer approval for the SI&V program. Typically, test specifications detail the tests that will be executed, the environment for executing the tests, and the pass criteria for the tests. Test procedure documents contain the detailed procedures. Often, the customer must approve these documents prior to the start of test execution.

Task 901 Define Operator and Facility Qualification Requirements

Identify the requirements for qualification of (1) the facilities where formal tests and demonstrations will be performed and (2) the personnel who will perform the formal tests or demonstrations. Consideration should be given to reporting needs and quality control.

Task 902 Develop Discipline Integration Plans

Develop the plan(s) for integrating the hardware, software, facilities, data and personnel of the system. The plan should contain the how, where and when of the integration process. The plan should include certification requirements for the integration process, facility and operators. There may be separate plans to cover different disciplines (e.g., electrical design, mechanical design, safety hazard analysis, etc.).

Task 903 Develop Discipline Verification Plans

Develop the plan(s) for verifying all specified requirements for the system. These requirements will be documented in formal specifications, Interface Control Documents (ICDs), and interface agreements. Requirements might also be contained in corporate or business unit standard documents (e.g. corporate logo standard). The plan should include the test philosophy, test plan, certification requirements for tests and test operators, and schedule and facility needs for testing and other verification. The plan should also include the approach for verification by analysis (e.g. simulators, stimulators, scenarios data) and by inspection (per your organization's standard IPQI process).

Task 904 Develop Assembly and Integration Procedures

Develop the procedures by which the system will be assembled and integrated. This may be documented in a Type D Process Specification.

Task 905 Design Integration Equipment, Facilities and Data

Design and/or specify the equipment and facilities needed to integrate the system. Design the data collection method. Develop data required to drive assembly and integration equipment.

Task 906 Design Verification Equipment, Facilities and Data

Design and/or specify the equipment and facilities needed to verify the system. Design the data collection method. Develop data required to simulate interfaces or system scenarios. Develop data required to drive test equipment.

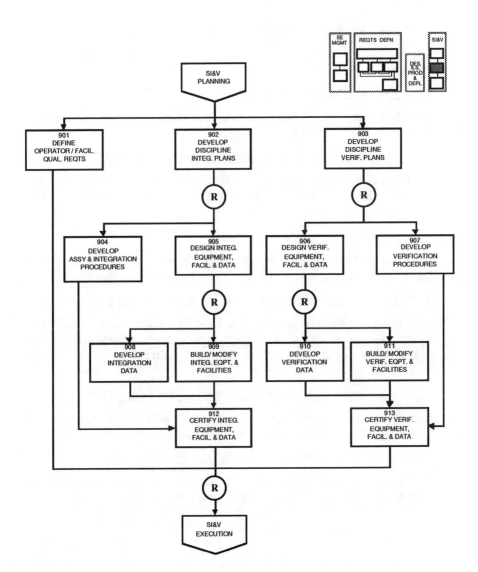

Figure 8-3 SI&V Development Activity

Task 907 Develop Verification Procedures

Develop the detailed, step-by-step verification procedures for the system.

Task 908 Develop Integration Data

Develop plans and drawings for the qualification and/or first article systems to be assembled.

Task 909 Build/Modify Integration Equipment and Facilities

Build/modify the equipment and facilities required to integrate the system. This may include building or procurement of new facilities or modification of existing facilities.

Task 910 Develop Verification Data

Develop the test cases for the system being integrated. These need to be based on the mission profile and expected system scenarios.

Task 911 Build/Modify Verification Equipment and Facilities

Build/modify the equipment and facilities required to verify the system. This may include building or procurement of new facilities or modification of existing facilities. Ensure calibration is current on all verification equipment to be used.

Task 912 Certify Integration Equipment, Facilities and Data

Certify assembly procedures, integration facility and operators to ensure they meet all applicable requirements, policies and procedures.

Task 913 Certify Verification Equipment, Facilities and Data

Certify verification procedures, test facility and operators to ensure they meet all applicable requirements, policies and procedures.

Table 8-2A   SI&V Development Input Checklist

Supplier	Input	Entry Criteria	Status
SE Mgmt	TPM Charts	Approved by SEM	
SE Reqts Defn	A/B/C/D/E Specifications	Approved by SEM	
SE Reqts Defn	Interface Control Documents (ICDs)	Approved by SEM	
SE Reqts Defn	System/Segment Design Document (SSDD)	Approved by SEM	
SE Reqts Defn	Test Requirements Specification (TRS)	Approved by SEM	
SE Reqts Defn	Technical Parameter (TP) Metrics	Inspected	
SE Reqts Defn	Technical Performance Measurement (TPM) Parameters	Approved by SEM	
SE Reqts Defn	Operational Concept Document (OCD)	Approved by SEM	
SE Reqts Defn	Mission Profile	Approved by SEM	
SE Reqts Defn	Measures of Effectiveness (MOEs)	Approved by SEM	
Design	Design Package	Inspected	
SW Dev Team	Software Development Plan (SDP)	Approved by SEM	
Deployment	Deployment Procedures	Inspected	
ILS	Technical Manuals	Inspected	
SI&V Planning	System Integration and Verification (SI&V) Plan	Inspected	
SI&V Planning	Requirements Verification Matrix (RVM)	Inspected	

Table 8-2B   SI&V Development Output Checklist

Customer	Output	Exit Criteria	Task #	Status
SI&V Exec, External Customer	Discipline Integration Plans	Inspected	902	
SI&V Exec, External Customer	Discipline Verification Plans	Inspected	903	
SI&V Exec	Integration Procedures	Inspected	904	
SI&V Exec	Verification Procedures	Inspected	907	
QA	Integration Equipment, Facility, & Data Certification Report	Approved by SEM	912	
QA	Verification Equipment, Facility, & Data Certification Report	Approved by SEM	913	

8.3 SI&V Execution Activity

Once the SI&V procedures, test environment, any special test software or hardware, and the product are ready, SI&V Execution Activity may begin (Figure 8-4). The primary focus of each verification step is to obtain adequate data to be able to ascertain how well the product performed its required functionality. Typically, *system integration* focuses on identifying and removing defects from the product while *system verification* focuses on demonstrating to the customer that the product is acceptable to them (by satisfying their requirements).

Tests may be performed using the Test, Analyze and Fix (TAAF) approach. Test failures may be recorded and analyzed using a Failure Reporting and Corrective Action System (FRACAS). Retests are performed as required.

This activity ends with the generation of a report (for each unique category of verification) that summarizes the results of the verification activities. The report usually includes Quality Assurance (QA) certification of the verification results.

Task 1001 Verify by Inspection

Verify by inspection to assure the requirements identified as "I" in the RVM are met. An inspection team, one of which is the developer, meet in a formal review to ensure that the item meets its requirements.

Task 1002 Verify by Analysis

Verify by analysis to assure the requirements identified as "A" in the RVM are met. It is very similar to the inspection technique in that no procedure is executed. Instead of simply reading the computer program or using error checklists, the participants use a small set of paper test cases.

Task 1003 Perform Process Qualification

Perform Process Qualification to demonstrate that a process exists and can be performed. This activity is associated with the "T" and "D" requirements identified in the RVM. In some instances, process qualification must precede facility and operator qualification. It is important to have manufacturing participate if they are involved in later execution steps.

Task 1004 Perform Facility and Operator Qualification

Perform facility qualification to demonstrate that the facility is in place and capable of supporting the integration and test activities. Perform operator qualification to show that the operators are ready (e.g. trained). Production Readiness Reviews, if applicable, should take place at this time.

Task 1005 Perform Incoming Test/Inspection

Perform incoming test/inspection to ensure that product is acceptable and meets the hand-off criteria before integration. The hand-off criteria should be established by the integration and test team and agreed to by the supplier prior to delivery.

Task 1006 Perform In-Process I&V

Perform in-process integration and verification to exercise the functionality of the product to detect and remove any defects that might be present at the various levels of integration. This involves the incremental assembly and checkout of subsystems until all the subsystems are integrated.

It verifies that the system functions under all of the conditions defined in the test plans/procedures by stressing the product to the limits of specified requirements. Once defects are identified and corrected, regression testing is necessary to assure that the changes have not broken the existing product.

Task 1007 Conduct Test Readiness Review (TRR)

Conduct Test Readiness Review to demonstrate that the integration and test team is ready to start the formal test/demonstration activities. This includes evaluating the integration test procedures, test results and open modification requests (MRs) to determine their impact on the formal test program. Results should be shared openly with customer emphasizing what functions have been tested thoroughly and what hasn't been adequately tested.

Task 1008 Verify by Formal Test/Demonstration

Verify by formal test or demonstration[40] the acceptability of the product. The requirements that need to be formally verified by this means will be indicated in the RVM by either a "T" for test or a "D" for demonstration. Acceptability is accomplished by executing all tests that are defined in the customer approved test specifications and procedures.

---

[40]A test usually involves some sort of instrumentation and collection of data. A demonstration usually does not involve instrumentation, but rather verifies compliance by mere observation of results. A demonstration "checklist" may be used to facilitate recording of demonstration results.

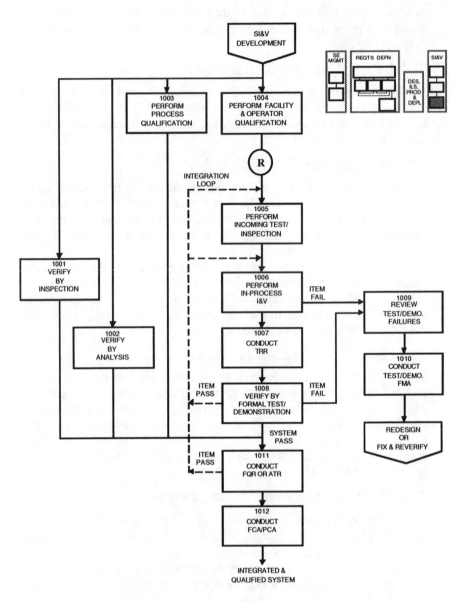

Figure 8-4 *SI&V Execution Activity*

Customer participation or observation may occur during these formal tests. The quality organization should also be involved with their participation indicated by signing off on all completed test data sheets.

Task 1009 Review Test/Demonstration Failures

Review test/demonstration failures in a fact-finding meeting to discuss and resolve any MRs generated during the in-process testing or formal test/demonstration. This is normally the first step in the corrective action process.

Task 1010 Conduct Test/Demonstration Failure Mode Analysis (FMA)

Conduct Test/Demonstration Failure Mode Analysis to determine the root cause of any MRs and to provide the corrective action required to resolve the issues.

Task 1011 Conduct Formal Qualification Review (FQR) or Acceptance Test Review (ATR)

Conduct Formal Qualification Review or Acceptance Test Review to verify that the actual performance of the product, as determined through test, complies with the Hardware Development Specifications, Software Requirements and all Interface Requirement Specifications.

Task 1012 Conduct Functional Configuration Audit (FCA) and/or Physical Configuration Audit (PCA)

Conduct Functional Configuration Audit and/or Physical Configuration Audit to validate that the development of a product has been completed satisfactorily. The FCA verifies that the product has achieved the performance and functional characteristics specified. The PCA is the final examination of the as-built version of the product in order to establish the product baseline. This includes a detailed audit of engineering drawings, specifications, technical data and test results for hardware and a detailed audit of design documentation listings and manuals for software. Where feasible, the FCA and PCA can be combined.

NOTE: A preliminary PCA is often done prior to, or immediately after, the Test Readiness Review (TRR). This ensures better configuration control since you will know exactly what you are testing.

Table 8-3A   SI&V Execution Input Checklist

Supplier	Input	Entry Criteria	Status
Design	Design Package	Inspected	
SW Dev Team	Software	Inspected	
Production	Hardware	Inspected, PCA	
Deployment	Deployment Procedures	Inspected	
ILS	Technical Manuals	Inspected	

Table 8-3B   SI&V Execution Output Checklist

Customer	Output	Exit Criteria	Task #	Status
Failure Review Board (FRB)	FRACAS Reports	Inspected	All	
SE Mgmt	Test and Evaluation Reports	Inspected	1001, 1002, 1008	
Ext Customer	Process Qualification Report	Appvd by SEM/ QA	1003	
Ext Customer	Facility/Operator Qualification Report	Appvd by SEM/ QA	1004	
CCB	Test/Demo Failure Reports	Appvd by SI&V CAM	1008	
Ext Customer	TRR Meeting Minutes	Appvd by SEM	1007	
Ext Customer	FMA Report	Appvd by SEM	1010	
Ext Customer	FQR/ATR Meeting Minutes	Appvd by SEM	1011	
Ext Customer	FCA/PCA Meeting Minutes	Appvd by SEM	1012	
PM, External Customer	Integrated & Qualified System	FCA, PCA	1012	

chapter nine

Process tailoring

"Fortune brings in some boats that are not steered."
—William Shakespeare

The process described in this book may be applied to all product development projects. In each application, this process should be tailored to the specific requirements of a particular project, project phase, or contractual structure. Care should be taken to eliminate tasks that add unnecessary costs, and that do not add value to the process or product. Tailoring takes the form of deletion, alteration or addition. Tailoring specific tasks requires definition of the depth of detail, level of effort required, and the data expected.

Thus, tailoring is performed to both breadth and depth. Tailoring in breadth of application is based on the project and project phase (e.g., types and number of systems impacted by the development of a new general application subsystem, the numbers and types of assessments, number and types of reviews).

Tailoring in depth involves decisions concerning the level of detail needed to generate and substantiate the outputs required to satisfy contractual objectives. The depth that the systems effort should take will vary from project to project in relationship to complexity, uncertainty, project urgency, and the willingness to accept risk. Selected and tailored requirements and task statements may be used by project managers in preparing solicitation documents and by offerors in response to a draft Request for Proposal.

9.1 Tailoring Considerations

The objectives of the contract effort and the inputs to the SE process depend on the breadth and depth of application. To assist in defining the depth of application and level of effort required, the following inputs should be identified for any application of the SE process:

a. The level of detail in system definition required from the contracted effort. For example, during conceptual investigations a complete functional decomposition of the system for each system alternative is not always necessary. However, sufficient depth is necessary to provide confidence in cost, schedule and performance objectives and related risk estimates. Different depths may be identified for areas in relationship to the application of new technologies.

b. The scenarios and mission to be examined for each primary life cycle function of the system.

c. A set of measures of effectiveness organized hierarchically. The relative importance of all metrics at the top level in the hierarchy should also be identified.

d. Known constraints and requirements for establishing constraints in areas where they are likely to exist but quantitative data are not available (or determine these internally).

e. The technology base and any limiting criteria on the use of technologies.

f. The factors essential to system success, including those factors related to major risk areas (e.g., budget, resources and threat).

9.2 General Guidance

The key elements of Systems Engineering as defined in Section 1.6 must usually be present in the tailored process. If one of the key elements is not used, caution must be taken to ensure that the risk of doing so is acceptable to the project.

Even in the unusual case in which a project does not believe it can use even a tailored version of this process, the principles defined in this book can certainly be applied to the project to reduce its risk and improve the quality of its products and processes.

It is the responsibility of the Process Champion to complete the tailoring for a particular project. See Appendix D for a description of the roles and responsibilities of the Process Champion.

To begin the definition of the project's SE process, programmatic (SOW, schedule, funding, etc.) and technical (functional, performance, etc.) requirements should be obtained. These requirements drive the definition of the SE subprocesses and their phasing in time. Also, these requirements guide the determination of the criticality of the system and its components. Next, identify the required tasks and the level of effort required for each. The final step in the tailoring process is the phasing of the process activities in terms of the tasks required. The project schedule (or SEDS) guides the task phasing.

The basic SE process described in this book is applicable to any development project (including new developments, modifications and product improvements), regardless of size or complexity. There are, however, some tasks that may require some specific tailoring.

For example, an unprecedented, new system development in concept exploration phase is not likely to require configuration management audits or formal change control mechanisms. However, conceptual exploration investigation of modifications to an existing or foreign developed system may need this type of activity (for example, to verify interface constraints).

TPM in concept exploration phase may be reduced to tracking critical technical objectives or decision metrics related to validated needs. A technology project may not require the execution of any TPM tasks although some top-level tracking of key success metrics is recommended.

Appendix C describes the typical acquisition life cycle for government projects. This Appendix can be used for guidance in tailoring phase-dependent activities and tasks.

9.3 In–Depth Considerations

The level of detail expected from the end-products of the technical effort must be identified, as this determines the depth to which the SE process must be executed. For example, functional analysis and synthesis should be conducted to a sufficiently detailed depth to identify areas of technical risk appropriate for consideration for the particular acquisition phase or effort.

The term "sufficiently detailed" is determined based on the objectives of the contracted effort. For example, sufficiently detailed in Concept Exploration and Definition phase is driven by the requirement (for unprecedented systems) to provide a conceptual solution to needs and to identify key technology requirements. During Engineering and Manufacturing Development (EMD) the amount of "sufficient detail" is all the way down to the bottom of the system definition where detailed design activities can start. Throughout acquisition, the level of detail may vary since the baseline system may be at one level of detail and product or process improvements or other modifications may be at a different level of detail.

9.4 SEMP Tailoring

The specific content requirements for the SEMP may vary for each application. Tailor the SEMP to define the specific content requirements for the intended application of the SE process. For example, if TPM were not implemented on a technology project, delete the requirements for this type of planning from the SEMP. If there are no technology insertion efforts being considered, delete those requirements.

9.5 Specialty Engineering and Functional Discipline Tasks

The tasks for specialty engineering (such as reliability, maintainability, safety, logistics support, training, disposal, etc.) and functional disciplines (such as mechanical, electrical, software, etc.) should be integrated into the SE process. Generally, the need (and resulting tailoring guidance) results from customer inputs and as a result of expected SE process results. The tasks identified represent some of the critical disciplinary considerations to be examined during development. As requirements are definitized, tasks from appropriate standardization documents (e.g., EIA 632) should be examined, selected and integrated into planned SE process activities.

Some tasks (e.g., ILS) are applicable throughout acquisition generally because support is always a factor. The tailoring of these tasks is expected to vary by application and project complexity. Depending on the specific project, some or even most of the tasks could be deleted. Also, the engineering specialties personnel needed to execute these tasks may be integral members (either as core or extended members) of the multi-disciplinary teams assigned to subsystem developments, commensurate with the system and acquisition phase.

9.6 Tailoring Documentation

The tailoring of the SE process should be documented for the benefit of all who must execute the tailored process. The worksheet in Table 9-1 may be used in the tailoring process. Sometimes the contract may specify how the SE process must be documented. Here are some suggestions for documenting the tailored process:

a. Explicit process and task descriptions and output document descriptions may be documented in the SEMP.

b. A redlined version of this book could be used where tailoring does not need to be formally documented.

c. Document templates could be developed as examples of the depth of detail required for a particular project.

d. The process could be modeled in a scheduling tool showing explicitly the tasks to be accomplished and the documents to be generated.

Table 9-1   *Tailoring Worksheet*

Task	Description	WBS Element	OBS Element	SEDS Task #
100	**SE PLANNING**			
101	Identify Corporate Resources			
102	Tailor SE Process			
103	Define Methodologies for System Analysis, Optimization & Development			
104	Develop Technology Insertion Approach			
105	Identify Corp. Procedures & Planning Baselines			
106	Assess Technical Program Risks & Issues			
107	Define TPM Parameters & Procedures			
108	Define Technical Review and Audit Plans			
109	Develop ILS, Prod, and Depl Approaches			
110	Develop DTC Objectives			
111	Develop SEMS, SEDS and WBS			
112	Define Staffing/Discipline Needs			
113	Define, Develop & Acquire SEDE and Tools			
114	Prepare Other Plans			
115	Prepare SE Management Plan (SEMP)			
116	Define Organizational Roles and Responsibilities			
117	Prepare Risk Management Plan			
118	Establish/Implement Risk Mgmt Program			
119	Support Project Management Process			
200	**SE CONTROL & INTEGRATION**			
201	Project Engineering			
202	Risk Management			
203	Technical Parameter (TP) Tracking			
204	Assess and Review Tech Program Progress			
205	Configuration Management			
206	Interface Management			
207	Data Management			
208	Technical Reviews and Audits			
209	Requirements Management			
210	Engineering Integration			
211	In-Process Quality Inspection (IPQI)			

Table 9-1 (cont'd)   Tailoring Worksheet

Task	Description	WBS Element	OBS Element	SEDS Task #
300	**REQUIREMENTS ANALYSIS**			
301	Collect Stakeholder Requirements			
302	Define System Mission/Objective			
303	Define System Scenarios			
304	Define System Boundary			
305	Define Environmental & Design Constraints			
306	Define Operations & Support Concept			
307	Define Measures of Effectiveness (MOEs)			
308	Define/Derive Functional & Perf Reqts			
309	Validate Requirements			
310	Integrate Requirements			
400	**FUNCTIONAL ANALYSIS & ALLOCATION**			
401	Define System States & Modes			
402	Define System Functions			
403	Define Functional Interfaces			
404	Define Performance Requirements & Allocate to Functions			
405	Analyze Performance & Scenarios			
406	Analyze Timing & Resources			
407	Analyze Failure Mode Effects and Criticality			
408	Define Fault Detection & Recovery Behavior			
409	Integrate Functions			
500	**SYNTHESIS**			
501	Assess Technology Alternatives			
502	Synthesize System Element Alternatives			
503	Allocate Functions to System Elements			
504	Allocate Constraints to System Elements			
505	Define Physical Interfaces			
506	Define Platform and Architecture			
507	Refine Work Breakdown Structure (WBS)			
508	Develop Life Cycle Techniques & Procedures			
509	Check Requirements Compliance			
510	Integrate System Elements			
511	Select Preferred Design			

Table 9-1 (cont'd) **Tailoring Worksheet**

Task	Description	WBS Element	OBS Element	SEDS Task #
600	**SYSTEM ANALYSIS & OPTIMIZATION**			
601	Develop System Models			
602	Perform System & Cost Effectiveness Analysis			
603	Risk Evaluation			
604	Trade Studies			
700	**REQUIREMENTS AND ARCHITECTURE DOCUMENTATION**			
701	Develop Document Approach			
702	Develop Detailed Document Outline			
703	Develop Text			
704	Develop Graphics			
705	Produce Document			
706	Deliver Document			
800	**SI&V PLANNING**			
801	Analyze Customer and Higher Level I&V Plans			
802	Develop RVM			
803	Define Integration Requirements			
804	Develop I&V Strategy/Approach			
805	Define Verification Requirements			
806	Define Facility Needs			
807	Define Staffing/Discipline Needs			
808	Define Data Needs			
809	Define Equipment (HW/SW) Needs			
810	Prepare System Integration & Verification (SI&V) Plan			

Table 9-1 (cont'd)   Tailoring Worksheet

Task	Description	WBS Element	OBS Element	SEDS Task #
900	**SI&V DEVELOPMENT**			
901	Define Operator and Facility Qualification Reqts			
902	Develop Discipline Integration Plans			
903	Develop Discipline Verification Plans			
904	Develop Assembly and Integration Procedures			
905	Design Integration Eqpt, Facilities and Data			
906	Design Verification Eqpt, Facilities and Data			
907	Develop Verification Procedures			
908	Develop Integration Data			
909	Build/Modify Integration Eqpt and Facilities			
910	Develop Verification Data			
911	Build/Modify Verification Eqpt and Facilities			
912	Certify Integration Eqpt, Facilities and Data			
913	Certify Verification Eqpt, Facilities and Data			
1000	**SI&V EXECUTION**			
1001	Verify by Inspection			
1002	Verify by Analysis			
1003	Perform Process Qualification			
1004	Perform Facility and Operator Qualification			
1005	Perform Incoming Test/Inspection			
1006	Perform In-Process I&V			
1007	Conduct Test Readiness Review (TRR)			
1008	Verify by Formal Test/Demonstration			
1009	Review Test/Demonstration Failures			
1010	Conduct Test/Demonstration Failure Mode Analysis (FMA)			
1011	Conduct Formal Qualification Review (FQR) or Acceptance Test Review (ATR)			
1012	Conduct Functional Configuration Audit (FCA) and/or Physical Configuration Audit (PCA)			

chapter ten

Systems engineering process support

"All things are ready, if our minds be so."
—William Shakespeare

This chapter provides information on training, tools, metrics, and tailoring guidelines that supports implementation of the SE process. Case studies are described that may assist the project teams in understanding different ways to apply this process.

It is assumed throughout this book that some sort of process improvement activity exists in the organization. There would be some sort of Process Management Team (PMT) or Process Action Team responsible for process improvements. It is also assumed that there is a Process Champion assigned to each project. The roles and responsibilities of the Process Champion are described in Appendix D. The Process Champion is not necessarily a full-time member of that project. This person could share this responsibility across several projects. Process support could be provided by a corporate level organization or could be local to that project. In some cases, such as for training, process support can be acquired from outside the organization.

Philip Crosby says, "Quality is free." This is true only if *net expenses* are considered. Both sides of the balance sheet must be addressed. Time and effort must usually be expended in improving your process, methods, tools and environment such that the overall increases in quality and productivity, and decreases in costs and cycle time will more than compensate for the expenses associated with process improvements.

10.1 Training

"I never let my schooling interfere with my education."
—Mark Twain

Reference books, textbooks, and other relevant documents are listed in the Bibliography Section of this book.

The following documents are recommended as reference material[41] for all systems engineers:

- EIA 632, Systems Engineering (industry standard)

- IEEE 1220, Standard for Application and Management of the Systems Engineering Process (industry standard)

- *System Requirements Analysis*, Grady

- *The Art of Systems Architecting*, Rechtin and Maier

- *Systems Engineering and Analysis*, Blanchard

The following documents, in addition to those listed above, are recommended as reference material for all systems engineering managers:

- *Systems Engineering Management*, Blanchard

- *Systems Engineering Management*, Lacy

Other sources of training are the Symposia sponsored by the International Council on Systems Engineering (INCOSE). The Proceedings from these Symposia are excellent source material for lessons learned, tools assessments, and process metrics. More information on INCOSE is provided in Section 10.5.

---

[41]All of these items are listed in the Bibliography.

10.2 Tools

"Give us the tools, and we will finish the job."
—Winston Churchill

Recommendations for use of specific tools is beyond the scope of this book. The performance and features provided by tools change too rapidly for any sort of assessment here to be valid for very long.

Many tools are described in Eisner's book called *Computer-Aided Systems Engineering*. There is a Tools and Automation Working Group in INCOSE that is an excellent resource for SE tools. (See Section 10.5 for more information about INCOSE.)

10.2.1 SE Management Tools

Whenever possible, use the same tools for SE Management as those being used by the Project Management process on a project.

10.2.2 Requirements and Architecture Definition Tools

Generally, some sort of CAE tool is needed for establishing and maintaining the Requirements/Decision Database. For a small project, this tool could be a simple relational database. For larger projects, usually a CAE tool specially designed to support systems engineering is required.

10.2.3 System Integration and Verification Tools

Often test engineers have special tools for their discipline. It is important that test planning use the same requirements traceability tools as used in Requirements and Architecture Definition Subprocess to avoid extra work.

10.3 Metrics

10.3.1 Purpose of Metrics

There are at least three reasons for collecting process metrics. First, you must be able to measure a process in order to improve it. Second, metrics provide project data for cost estimating and for planning the required activities and schedule intervals. Third, metrics provide a benchmark against which we can compare our performance to other projects and companies. A generic, standard process will more likely be adopted for use if we have metrics to prove actual benefits.

A metrics guidebook is available from INCOSE: "Metric Guidebook for Integrated Systems and Product Development." See Section 10.5 for information on how to contact INCOSE.

10.3.2 Metrics and the SE Process Champion

The SE Process Champions selected by project and functional management are responsible for collecting metrics data and reporting them to the SE PMT. The roles and responsibilities of the SE Process Champion are defined in Appendix D.

Collect this metrics data at product development checkpoints (or decision points) and report them to the SE PMT. A project level SE PMT may be formed to facilitate the collection of metrics data and tailoring of the generic process for each project.

10.3.3 Metrics Data Collection

The SE PMT may require that a set of metrics data[42] be collected and reported by each project implementing this process. Metrics may be derived from this set of data and these metrics will provide information to the SE PMT for measuring effectiveness and efficiency of the SE Process.

By continually monitoring these metrics, the PMT will: (1) provide oversight and guidance to lower tier PMTs and practitioners implementing their processes, and (2) continually improve the generic SE Process. Corresponding metrics for cost, performance, productivity

---

[42]A distinction is made here between *metrics data* and the *metrics* per se. *Metrics data* is the raw information that is used to derive the *metrics*. *Metrics data* is usually only useful to the metrics analyst who will analyze the data for trends and anomalies, and then calculate the relevant metrics. *Metrics* themselves can be used by various users such as those shown in Section 10.3.4.

and interval may be derived and reported to executive management by the SE PMT.

Metrics data may be collected by various means. Some of the data may be collected from the project through the Process Champion. Other data may be collected from existing sources such as the financial management system (cost and schedule) and survey data (customer satisfaction).

10.3.4 Types of Metrics

Metrics can be categorized as shown in Table 10-1. This table also shows some of the uses for each type of metric. There is a broad range of users of metrics. The different classes of metric users often have very different needs. Some need metrics almost on a daily basis to ensure the efficient running of a project or task. Others need metrics on a quarterly or yearly basis to determine if corporate goals are being met. And then others may need metrics somewhere in-between, perhaps to improve the overall SE process or to measure near-term staffing needs.

There is some additional information that is useful to the metrics analyst:

a) Normalization factors,

b) Quality correlation factors, and

c) Cost estimating relationships.

Normalization factors are used to normalize the raw data or the metrics itself so they can be reasonably compared between projects, companies, etc. Examples are project type (commercial vs. military), project phase (concept exploration vs. full scale development), system precedence (new start vs. upgrade), and product complexity (few vs. many requirements).

Quality correlation factors are information to determine what factors are responsible for increasing or decreasing quality of an effort. It might be determined through analysis of metrics data that use of a certain tool on some projects consistently increased quality and productivity by 20 percent. This would be a necessary correlation to find in order to justify investment in the acquisition, training and support costs for that tool. Or there may be a case where a particular methodology is increasing the time to complete a task without a corresponding increase in either quality or productivity. Appropriate action would be required to either improve or eliminate that methodology.

Table 10-1   Basic Metrics Types

Basic Metric Type	Aliases	Typical Uses
Customer Satisfaction	Quality, Value, Value-Added	• Set priorities so as to maximize award fees and profit • Keep "in touch" with the customer's needs • Determine "value-added" for downstream internal customers
Performance	Defect Density, Quality, Functionality	• Determine reliability of delivered product • Determine if defects are being found early enough in a project to avoid costly rework downstream (Note: For systems engineering, this metric is very important since the defects introduced during the requirements definition phase will have the greatest cost/schedule impact for a project.) • Used to find the root cause of defects to determine where to improve the process
Productivity	Efficiency	• Determine efficiency of staff or physical resources (facilities, equipment, tools) in performing a given task • Used to justify investment in new methods, tools, and facilities
Cost	Staff Months, Price	• Benchmark costs against other business units and companies • Determine if meeting cost and expenditure goals • Compare budgeted versus actual costs • Develop quote for a project or task
Interval	Cycle Time, Timeliness, Schedule	• Benchmark task intervals against other business units and companies • Compare budgeted versus actual time • Develop schedules for a project or task
Process Maturity	Capability Maturity	• Determine "best in class" for process, methods, tools and environment • Set objectives for process maturity • Benchmark against other business units and companies
Process Compliance	Policy Compliance	• Determine degree of success in deploying a process and its associated support material (training, tools, documentation) • Determine effectiveness of policy and procedures
Number of Process Improvement Opportunities	Number of Customer Suggestions	• Identify level of process improvement activity • Determine level of process quality awareness • Determine morale of organization
Product Maturity	Technical Maturity, Technical Progress	• Determine technical risk levels • Determine technology maturity • Determine viability of technology transition plans • Assess readiness for a major milestone (e.g. # TBDs/TBRs) • Assess completion of a major milestone (e.g. # action items)

Cost estimating relationships (CER) are used to perform a parametric analysis of costs. The CERs will allow a cost engineer to use cost data from another project, especially if the hardware or software on that other project is not similar enough to justify using a simpler similarity analysis. The use of CERs can greatly increase the accuracy of a cost quote (especially in concept and study phases) and decrease the actual cost of developing the quote. Examples of CER factors are number of interfaces, complexity of interfaces, type of interfaces, platform (shore, ship, underwater), reliability levels, and support factors (local vs. depot, customer vs. contractor).

10.4 Case Studies and Examples

The case studies below are intended to assist the project teams in more effectively applying the Systems Engineering process.

10.4.1 On the Use of Tailored SEMPs on Small Projects

Small projects or even larger ones which are mostly mature but which can benefit from the introduction of systems engineering disciplines might consider the following approaches to a technical management plan.

A contract may not call for a SEMP; however, all projects need a plan for managing technical tasks to maintain internal discipline, produce quality results, and contain costs. The goal in such cases is not necessarily to implement a redefined subset or to partially implement each task of the full SEMP. Rather, the systems engineering manager should identify the elements of the project which need systems engineering attention and tailor a plan from both the more comprehensive SEMP format and the available processes for systems management of the project. Document the plan in a few pages and get buy-in from the project and technical organizations.

In general, every project ought to have a schedule (produced on one of the popular project management programs such as TimeLine, or Microsoft Project), a requirements traceability matrix which includes tests (for small enough projects use a commercial database such as dBase, Paradox, or Access), configuration management, and a modification request system.

Documents produced (if not specified by the contract) should include as a minimum the plan itself, the schedule (including resources, costs, and dependencies), a requirements specification, the requirements traceability matrix (which can include system tests), a configuration item list for hardware and software in sufficient detail

that all personnel know what they are producing, and a set of
modification requests which undergo periodic review.

Other elements of processes can be used and specified if they
benefit the project. Metrics should be kept on, as a minimum, cost,
interval, productivity, and performance. The following are two
examples of this approach.

10.4.1.1 DCSS Example

The Digital Conferencing and Switching System (DCSS) can be
characterized as an existing system with a specially designed hardware
switch and very complex conferencing software both within the switch
and in PC controllers. It is a commercial grade product which
undergoes frequent updates to facilities and features. The system is
deployed in many sites, with customers who stress particular
conferencing capabilities. The design started out a number of years
ago as a project for a particular customer and did not develop a
hierarchy of specifications and requirements.

Accordingly, additions and changes must be made in a disciplined
way to avoid serious effects on customers, many of whom are using
the system for revenue production.

As the system has grown and found success in the marketplace, it
has shown the need for interjection of current processes and a technical
management plan which would achieve substantial benefits at
relatively low cost. The plan for this project extracted and tailored
relevant portions of the system design process and SEMP. The plan
established a front end process for introducing new features based on a
simplified version of the No. 5 ESS Feature Definition and Assessment
Form. It uses a trouble ticket and modification request (MR) process
review which flows customer inquiries and problems into software and
hardware changes.

It also captures functional and performance requirements in a
dBase file which includes test procedures and results, as well as a trace
to generated MRs. The MR tracking system is a simple program used
by DCSS for years. A detailed schedule is developed for all aspects of
development and includes specific personnel assignments as well as
loaded rates in order to estimate and control costs.

Other processes and disciplines are introduced as needed. For
example, as a risk assessment and mitigation task, the early stages of
the SE management efforts included Pareto analyses of both trouble
tickets and factory repairs to identify the dominant problems being
experienced by customers and failures experienced by the equipment.
These were then attacked in order of priority to improve customer
satisfaction and significantly reduce the repair costs.

All the above methods and processes were described in a plan, consisting of a few pages, which outlined the SE efforts to be undertaken and identified the documents which would be produced. The improvements implemented accomplished on the order of 60 to 70 percent of the totality of efforts which could have been usefully introduced.

10.4.1.2 SOSUS Consolidation Project Example

The SOSUS Consolidation Project is reaching its conclusion and the project is going into a largely maintenance mode. As the customer plans consolidation of the sites into fewer entities, changes are introduced into the system hardware and software. At the same time funds are rapidly decreasing and personnel must move off the project. This situation creates pockets of vulnerability which must be addressed and resolved in order to successfully complete the project.

The remainder of the consolidation project has been organized as an Integrated Product Team (IPT), with products identified as the consolidated sites or as communications upgrades to sites. Efforts of Systems Engineering have been concentrated on detailed scheduling of remaining tasks where needed, monthly requirements reviews for upcoming site consolidations, processes for material management, a process for control of software modification requests, and management of the sequence and efficiency of activities necessary to define configurations, produce application schematics, and plan and execute installations. The areas of focus and approaches are documented in a plan which consists of a few pages, plus schedules and/or process documents for each area of focus.

10.4.1.3 Tailored SEMP Conclusions

The above examples illustrate that each project has its own unique needs which might not have been anticipated by a generic simplified SEMP. There are, of course, the common elements cited above. In both cases it may be of interest to note that introduction of the processes was bought into by the project management only by a fairly rapid demonstration of the benefits of the selected processes. In other words, systems engineering had to concentrate on demonstrating significant benefits for the funds spent, as opposed to just following all the processes which could have been used. Such a situation might be expected to be the case for a lot of small or mature projects with limited systems engineering budgets.

10.4.2 SE as Applied to Life Prediction and Testing

The systems engineering (SE) process was successfully tailored and deployed on a hardware-intensive task that initially was not considered a "typical" SE work package. On the STS project, the Life Prediction and Testing Task required a substantial amount of Life testing whereby precision test fixtures were designed, fabricated, and assembled into integrated Life test sets. The successful execution of this technical task depended not only on satisfying performance objectives but also on the control of cost and schedule elements. The tailored SE process proved to provide an excellent framework from which the Task's activities could be performed in a methodical manner.

At the commencement of the project, the Task's activities were mapped to the SE Process. The standard Process provided an excellent template to perform the activities that are routinely incorporated when systematically attacking and solving a problem. Such activities to cite a few included verifying, defining, developing, specifying, and integrating requirements; identifying, assessing, managing, and mitigating risk areas; defining and analyzing functions, components, interfaces, TPMs, and MOEs; etc. These tailored SE activities pertained to the performance, cost, and schedule elements of the Task.

The following Table captures the actual amount of time that the Life Task Team spent on each SE activity:

SE Task #	SE Task	Time (wks)
106	Assess Technical Risks	6.0
107	Define TPM Parameters	0.4
202	Risk Management	5.4
203	Technical Parameter (TP) Tracking	0.2
305	Define Environmental & Design Constraints	2.4
307	Define Measures of Effectiveness (MOEs)	0.2
308	Define/Derive Performance Requirements	1.0
309	Verify Requirements	0.5
310	Integrate Requirements	1.6
401	Define System States and Modes	0.4
403	Define Functional Interfaces	1.0
404	Allocate Performance Requirements to Function	1.0

407	Analyze FME&C	1.0
409	Integrate Functions	0.2
	TOTAL=	21.3

Note that there was a considerable amount of overlap in effort between several of the above SE tasks (e.g., between Tasks 305, 308, 309 and 310; and between Tasks 403, 404 and 409). This total time of 21 weeks dedicated to SE activities represents approximately 20% of the total time committed to the Life Prediction and Testing Task.

Technical, cost, and schedule analyses, decisions, and results were captured in a SE notebook which was not deliverable to the customer. This informal SE document was maintained by all the Life Prediction and Testing Task team members. To this end, emphasis was placed on capturing, flowing, and reviewing information and requirements in real time among the entire STS Team. Although this was generally accomplished fairly informally via meetings and handouts, the Team was always kept up to date resulting in greater productivity.

In sum, the tailored SE Process provided an excellent framework from which the Task's activities were successfully executed in a structured manner. It is worthwhile to note that this life effort was very well received by the customer and was key in the win of the STS award.

10.5 International Council On Systems Engineering (INCOSE)

The International Council On Systems Engineering (INCOSE) is an international organization formed to develop and enhance multi-disciplinary system development under the title of Systems Engineering. INCOSE can help you become a leader in your company by applying Systems Engineering in solving your company's problems.

INCOSE was created to:

1. Foster the definition, understanding and practice of world class systems engineering in industry, academia, and government.

2. Provide a focal point for dissemination of systems engineering knowledge.

3. Promote collaboration in systems engineering education and research.

4. Assure the existence of professional standards for integrity in the practice of systems engineering.

10.5.1 INCOSE Working Groups

INCOSE has many working groups exploring different aspects of systems engineering. More groups are formed as the need arises.

Working Group	Activities
Best Practices	Identifies examples of "World Class" and "Preferred Future" practices.
Policy Review	Develops and coordinates INCOSE position statements on proposed policy.
Process Description	Compares and contrasts processes from various sources.
Tools	Collects and disseminates information on tools to support Systems Engineering.
Metrics	Develops and calibrates methods for SE process measures.
Requirements Management	Identifies effective approaches for defining and integrating requirements.
Principles	Identifies and documents Systems Engineering techniques and heuristics.
Capability Assessment	Development of a maturity model to assess an organization's SE capability.
Resource Management	Areas of interest include, but are not limited to, applying SE to energy, transportation, agriculture, municipal planning, environment planning, and forestry management.
Risk Management	Develops risk management principles and practices. Defines best practices.
Commercial Practices	Seeks to identify and disseminate information and guidance on the practice of SE in many and varied commercial sectors.
Concurrent Engineering	Explores the role of SE in concurrent engineering.

10.5.2 INCOSE Membership Benefits

INCOSE is one of the only professional associations dedicated entirely to systems engineering. There are many benefits to membership in INCOSE:

a) Technical Symposia and Papers

b) National Journal

c) Seminars

d) Tutorials

e) Local Chapter Activities

f) Working Groups

g) Technical Committees

h) Leadership Opportunities

i) SE Tool Demonstrations and Lessons Learned

j) Networking with fellow system engineers

10.5.3 Further Information

INCOSE can be reached by various means:

- http://www.incose.org
- incose@halcyon.com
- (800)366-1164

chapter eleven

Programmatic application

"We must all hang together, or most assuredly we shall *hang separately.*"

—Benjamin Franklin

The Systems Engineering Process cannot be performed well in a vacuum. It requires the right people, the right organization for those people, and the right functional design disciplines involved (e.g., software, mechanical, circuit card, ASIC/FPGA, microwave, interconnect assemblies, and test equipment).[43] It also requires the proper timing with regard to when the different "levels" of the system architecture are defined and when the related functional design disciplines are brought to bear.

This chapter describes the programmatic framework within which the Systems Engineering Process is performed. This framework involves the system breakdown structure (or product hierarchy), the organizational breakdown structure (or project teaming chart), and the project schedule.

11.1 System Architecture

The Systems Engineering Process described in this book should be applied recursively at each level of the system's product hierarchy. The system is defined down to the point where the requisite functional

---

[43] Of course, there can be other disciplines involved, like contracting, procurement, manufacturing, deployment, and specialty engineering (reliability, maintainability, safety, human factors, logistics support, training, environmental, etc.).

design disciplines (FDDs) can perform their specialized tasks more or less independently.

This "decoupling" is necessary to allow parallel developments to occur with minimal effect on the other products. This "divide and conquer" approach often leads to the shortest cycle time since many activities can be going on at the same time. This requires good systems engineering (the process, not the people) so that any changes on one product will be less likely to affect the others. However, this approach depends on an up-front investment of time and people. If the downstream stakeholders are not involved up-front in the Systems Engineering Process, there is a good chance that the system's products will not be sufficiently decoupled from each other and important requirements will be missed.

Figure 11-1 shows a notional system architecture. In this example, the Systems Engineering Process will be performed seven times, once for each shaded block[44] on the product hierarchy diagram. The levels in the hierarchy can be thought of as "levels of development." Notice that a FDD item[45] can be required at any level of the product hierarchy. For example, there may be test equipment required at the system level to test the functionality of the system after the two subsystems are integrated together. This test equipment may be unique for the whole project, or it may have some common features with lower-level test equipment.

The tasks of the Test Equipment FDD may be performed early in the project schedule to design, analyze, or model such test equipment as needed in order to validate assumptions, characterize performance parameters, assess cost and schedule issues, identify risks, etc. In fact, the Test Equipment FDD tasks may need to provide data to the Systems Engineering Process tasks for such things as trade studies, technology assessments, risk assessments, and system evaluation. These same type of comments will apply to any other FDD that is necessary to the success of that system development.

---

[44] The shaded blocks in the product hierarchy are called here "System Items" to distinguish them from the FDD items. The distinction is that System Items involve more than one functional design discipline. The development of FDD items will, of course, involve other disciplines, but the primary effort is by the engineers from that particular FDD.

[45] A FDD item is defined as a product or portion of a product that can be designed wholly by one of the several primary Functional Design Disciplines (FDD), for example: software development, mechanical assembly, circuit card assembly (CCA), ASIC/FPGA development, microwave development, interconnect assembly, and test equipment. Of course, these are not the only possible FDDs. Each organization, company and industry will need to define their own particular FDDs.

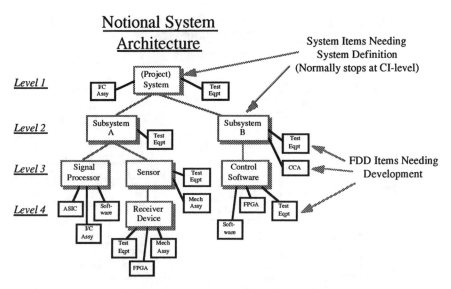

Figure 11-1 *Relationship Between System Definition "Building Blocks" and System Architecture*

11.2 Project Organization

Many problems occur on a project due to confusion about roles and responsibilities. Often there is confusion about who "owns" certain requirements, especially those that lie at interfaces between system products. To avoid many of these problems, it is often best to organize a project around the system architecture. Figure 11-2 shows a notional project structure that is mapped from the notional system architecture shown in the previous section. Notice that for every block in product hierarchy there is a corresponding team in the project structure. This offers clear delineation of responsibilities between the teams and the products.

The System Team is responsible for the project's system.[46] Each subordinate product has its own Integrated Product Team (IPT). The System Team and IPTs are mainly focused on the technical issues while the Leadership Team has a broader focus on the programmatic issues such as customer interface, marketing, business development, project strategy, corporate alliances, resource planning (staffing and facilities), contracting, etc.

---

[46] A project may actually be developing a subsystem, assembly, or lower level product in the overall system hierarchy, but for the purpose of this Chapter a "system" is defined as the top-level product that the project is responsible for developing.

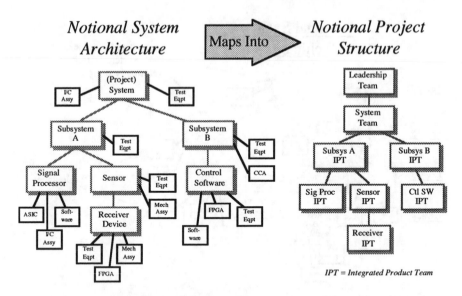

Figure 11-2   *Relationship Between System Architecture and Project Organization*

Smaller projects may choose to combine the System Team and the Leadership Team, but it is often wise to keep them separate for various reasons. The System Team is often emotionally attached to a particular approach whereas the Leadership Team can often take a more objective stance. The Leadership Team should avoid becoming too involved in the technical issues since this in some ways affects their objectivity.

The System Team and each IPT should be fully responsible for the corresponding FDD Items associated with their product. Each team should have the proper representatives from the relevant FDD and other appropriate specialties. The IPTs will not necessarily be established at the beginning of a project since the system architecture may not be known at that time. However, once an IPT is established usually that IPT is responsible until the product ships out the door and sometimes even for field acceptance.

All product-oriented teams should have a limited number of "core team" members. The ideal number for this core team is seven plus or minus two.[47] The core team members are usually designated for the

---

[47] This "ideal" team member count is based on several studies over the years in efficiency and productivity of groups of people. As the number of people in a group rises, the amount of effort in communicating with the other members increases faster. Hence, there comes a point when it becomes less efficient to add people to a group. Perhaps one of the original studies on this subject is documented in "The Magical Number Seven Plus or Minus Two: Some Limits on our

life of the team. The "resource" team members will represent other disciplines (technical and non-technical) that are not necessarily needed on a continuous basis and are assigned roles as appropriate.

The core team should decide when the "specialists"[48] are needed and should define the timing and level of their effort required. The System Team and Leadership Team should provide guidance to the IPTs with regard to the proper use of the specialists.

The Leadership Team and the System Team usually need to facilitate the formation of IPTs by recruiting and assigning the initial core team members. The initial core team can then determine what other core team members may be needed.

It is very important that the charter for each team be clearly established. One of the major failures of teams is caused by misunderstanding of roles and responsibilities. The roles, responsibilities and authority (RRA) of team members need to be established and agreed to by the team members. All *stakeholders* need to have a clearly identified *advocate* on each team.[49]

This project structure is not mandatory since often a project has other constraints that lead to different organizational structures. Many projects organize to line up with how their customer organizes. However, it is usually best to organize, whenever possible, consistent with the product hierarchy.[50]

11.3 Cross-Project Issues

There are times when the teams shown in Figure 11-2 cannot address some of the issues a project faces, especially when an issue involves multiple products or projects. Figure 11-3 shows various Cross-Product/Cross-Project Teams (CPTs) and how they relate to the product-oriented teams. Examples of CPT issues are: design-to-cost, six sigma, cycle time, product reuse, requirements management, configuration management, information management, quality assurance, and interface definition.

Capability for Processing Information" by George Miller (*Psychology Review*, 1956, Vol. 63, pp. 81-92).

[48] These "specialists" are not necessarily members of a specialty engineering discipline. In fact, there are occasions when one or more of the core team members will be a specialty engineer representing such disciplines as safety, reliability, maintainability, etc. Often the System Team will have at least one specialty engineer as a core team member.

[49] This is especially true for specialty engineering disciplines. There are rarely enough specialty engineers to staff all of the teams that are affected by what they do.

[50] Sometimes it can even benefit the customer to reorganize around the system architecture. This enhances communication and helps to avoid misunderstandings.

There are three types of relationships between a CPT and a regular project team: subordinate, superordinate, and informational. A CPT could be formed by an IPT to address a particular issue such as defining an interface between that product and some other item. That CPT would then be subordinate to the IPT that formed it. Or an IPT could be under obligation to follow the direction from a higher level team. The CPT may also have informational links to other product teams or even other projects or groups within your company or organization, whoever might be affected by the decisions of the CPT.

Often a CPT might be established to address FDD issues across the project. For example, the software development engineers may need to work together toward common procedures, tools, terminology, reusable components, etc. This is a convenient vehicle for "carrying the torch" of a particular discipline since often all of the other members of an IPT are from a different discipline than theirs; there is often a need to be able to share thoughts and ideas with other members of their discipline.

At the project level, the Leadership Team may form CPTs to address cross-project issues like product reuse, quality assurance, etc. The CPTs could be either ad hoc (temporary) or established for the duration of the project.

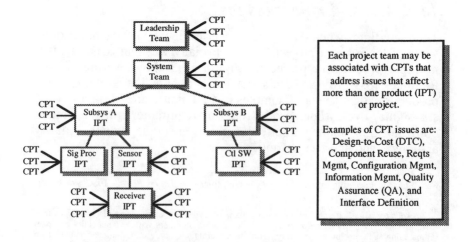

IPT = Integrated Product Team

CPT = Cross-Product/Project Team

Figure 11-3 *Addressing Cross-Product and Cross-Project Issues*

11.4 Time Dependencies

The Systems Engineering Process may be applied in two stages for each System Item in the system's product hierarchy: Conceptualize and Characterize. This section will focus on the time aspects of these two stages. These two stages are used for illustration purposes only. They have been found to be useful in certain applications, but each company or organization will need to determine the best manner in which to apply the Systems Engineering Process to its product development process.

The Conceptualize stage will often be a fraction of the effort compared to the Characterize stage (e.g., one-tenth the effort).[51] The main objective of the Conceptualize stage is to scan the horizon, to understand the "lay of the land," to identify the "gotchas" that will trip up the project. Then, during the Characterize stage, the focus should be on the higher risk elements of the product. Conceptualizing will lay the foundation for Characterization, thereby reducing the overall risk for the project.

It may appear that having two stages in a process where each stage has identical process tasks will increase cycle time rather than decrease it. This is one of the mysteries of concurrent engineering.[52]

11.4.1 Relationship to Schedule

Using this approach, cycle time can be reduced since effort will not be wasted on "low reward" tasks. The recursive nature of these two stages is equivalent to "peeling back the layers of an onion." Figure 11-4 shows how the two stages might appear on a time-based project schedule. Notice that the Conceptualize stage of the second-tier products can (and may need to) start *prior to* the finish of Conceptualize at the first tier. Also, the Characterize stage for the second-tier product can (and may need to) start *before* the Characterize stage of the first tier. The planning and design activities for the FDD

---

[51] The actual ratio will depend on many factors. One of the greatest factors is *degree of precedence*. If a problem is *highly precedented* (i.e., something the company, the project, and the people have routinely done before), then the ratio could be 20 or 30 to 1. However, if the problem is *highly unprecedented*, it may be wise to spend more effort up-front in the Conceptualize stage, perhaps a ratio of 3 to 1 (i.e., 25 percent of the total effort for that product would be spent doing Conceptualization).

[52] In a sense the two stages correspond to the popular phrase—"plan the work, then work the plan." This might be a loose analogy, but it does convey some of the reason for the need to conceptualize prior to characterizing the opportunity and its solution(s). You must have some sort of concept of what the problem is and some sense of the possible solutions before you can plan the work to fully develop the solution. Also, you need to know and understand the pitfalls and the major risks; conceptualizing first allows you to identify the pitfalls and risks without getting too quickly bogged down in the details of designing solutions.

items can start whenever the associated IPT deems it necessary based on a careful assessment of overall project risks. Starting too late increases cycle time, and starting too early leads to unacceptable risk of rework or misapplied resources.

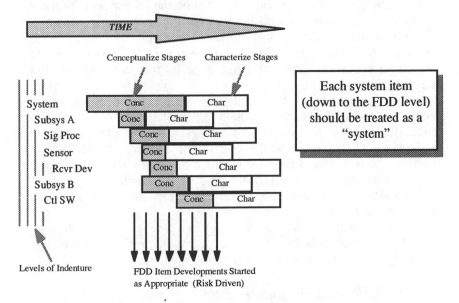

Figure 11-4   *Time Dependencies of the Conceptualize and Characterize Stages*

Even though there appears to be a "clean break" between the two stages, stages of any process are never a clean cutover in practice. There are always exceptions and circumstances that warrant some sort of overlap. This should be understood and decided upon in the Systems Engineering Planning tasks as to how much overlap is appropriate.

11.4.2 Time Gaps Between Stages

Figure 11-5 shows how there might be a gap in time between the two stages for a particular product. There are several reasons this might happen. A higher level Characterize activity may depend on a lower level product concept. Or the same people that would be doing Characterization at the higher level may be involved in Conceptualizing of lower level products. It is the responsibility of the System Team and Leadership Team to ensure that the time gaps between Systems Engineering Process tasks at any level do not adversely affect the project in terms of cost, schedule or risk.

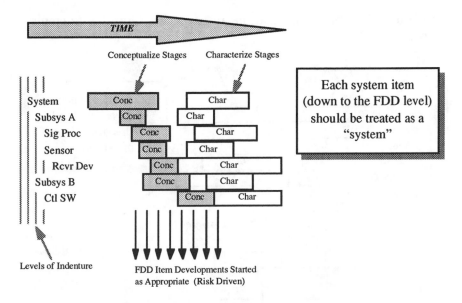

Figure 11-5 *Time Gaps Between Stages*

11.4.3 Contractual Situation

Sometimes the time gaps between stages have to do with contractual matters like RFPs and contract awards. Figure 11-6 shows a situation where some of the Conceptualization tasks are performed prior to receipt of the RFP and continue after receipt on up to contract award.[53] Contract award may be the time that Characterization tasks are first performed for a particular project.

In the example shown, there is a prime contractor responsible for the System and three separate subcontractors for Subsystem A. Each subcontractor implements his Conceptualize stage in a different manner with different timing. Also, notice that the third subcontractor has no activities for a receiver device since his particular sensor assembly either does not require a receiver device or perhaps the receiver device is off-the-shelf and therefore requires no development. Two of the subcontractors are eventually awarded the contract as a risk mitigation approach by the prime contractor.

The scenarios presented in this Chapter are examples only. Each project is unique and will need to construct its own project structures and engineering schedules. The principles espoused in this Chapter

---

[53] It is often prudent to continue performing some of the Systems Engineering Process tasks after submittal of the proposal and prior to contract award. Of course, this is dependent on funding and resource constraints.

should help your organization in implementing the Systems Engineering Process.

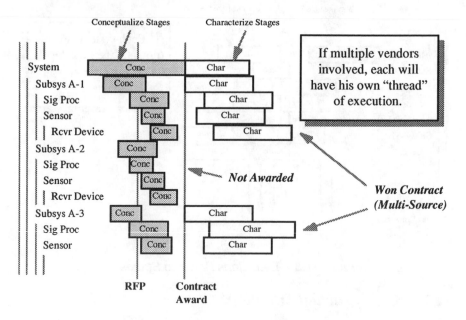

Figure 11-6 *Contract Award Scenario*

Glossary

"Novelty is mistaken for progress."

—Frank Lloyd Wright

A Spec   Common name for the <u>System Specification</u> or <u>Segment Specification</u> as defined by MIL-STD-961. See further definition below.

Acceptance   The process of proving that a product, process or material is acceptable to the customer. Acceptance may be done in the factory or on site. Usually the acceptance requirements are documented in the C/D/E specifications.

Also, the act of an authorized representative of the government by which the government, for itself, or as an agent of another, assumes ownership of existing identified supplies tendered, or approves specific services rendered, as partial or complete performance of the contract on the part of the contractor.

Allocated Baseline   The initially approved documentation describing a Configuration Item's (CI) functional and interface characteristics that are allocated from those of a higher level CI; interface requirements with other CIs; design constraints; and verification required to demonstrate the achievement of specified functional and interface characteristics. Consists of the Type B specifications that define functional requirements for each CI. Normally established at the Preliminary Design Review (PDR), but no later than the Critical Design Review (CDR).

Architecture   The highest-level concept of a system in its environment. An
                          architectural description is a model—document, product or other
                          artifact—to communicate and record a system's architecture. An
                          architectural description conveys a set of views each of which depicts
                          the system by describing domain concerns.

B Spec   Common name for the <u>Development Specification</u> as defined by
                          MIL-STD-961. This specification type has several subtypes. B1 is
                          for a prime item. B2 is for a critical item. B3 is for a non-complex
                          item. B4 is for a facility or ship modification. B5 is equivalent to a
                          Software Requirements Specification and its associated Interface
                          Requirements Specifications. A <u>Functional Specification</u> is a special
                          version of a B Spec which lists the requirements for a particular
                          functional discipline. See further definition below.

C Spec   Common name for the <u>Product Specification</u> as defined by MIL-STD-
                          961. This specification type has several subtypes. C1 is for a prime
                          item. C2 is for a critical item. The C1/C2 specs have further
                          subtypes of C1a/C2a for Product Function (form, fit and function
                          only) and C1b/C2b for Product Fabrication. C3 is for a non-complex
                          item fabrication. C4 is for an inventory item. (A C4 spec is used for
                          notifying a government agency that a pre-existing item will be used
                          on a particular system.) C5 is equivalent to a Software Product
                          Specification. See further definition below.

Chief Systems   The CSE is the lead engineer and chief architect for the entire system.
Engineer   This may be the same person as the Systems Engineering Manager
(CSE)   (SEM).

Concept   The government acquisition phase where system concepts of
Exploration   operation and support are developed and explored. One of these
and Definition   concepts is chosen for further study and definition during the
                          Demonstration and Validation phase. Sometimes called Phase 0.

Concurrent   A systematic approach to the development of products and their
Engineering   related processes that considers all aspects of the product life cycle
(CE)   from the outset, including: technical performance; cost; schedule;
                          producibility; testability; reliability; operability; maintainability;
                          affordability.

Configuration   An aggregation of hardware/software, or any of its discrete portions,
Item (CI)   which satisfies an end use function, and is designated by the
                          Design/Development organization or the customer for configuration
                          management. CIs may vary widely in complexity, size and type.

Configuration Management (CM)	A discipline of applying technical and administrative direction and surveillance to (a) identify and document the functional and physical characteristics of a configuration item, (b) control changes to those characteristics, and (c) record and report change processing and implementation status.
Cost Account Manager	A person responsible for the cost and schedule performance of a work package associated with a Work Breakdown Structure (WBS).
Cost and Operational Effectiveness Analysis (COEA)	An analysis of the costs and operational effectiveness of alternative materiel systems to meet a mission need and the associated project for acquiring each alternative.
Cost Breakdown Structure (CBS)	In accomplishing a life-cycle cost analysis, one needs to develop a CBS, or cost tree, to facilitate the initial allocation of costs (top-down) and the subsequent collection of costs on a functional basis (bottom-up). The CBS must include the consideration of all costs and is intended to aid in providing overall cost visibility.
Critical Design Review (CDR)	This review is conducted for each configuration item when detail design is essentially complete. The purpose of this review is to: (a) determine that the detail design of the CI under review satisfies performance and engineering specialty requirements of the hardware configuration item (HWCI) development specifications, (b) establish the detail design compatibility among the CI and other items of equipment, facilities, software, and personnel, (c) assess configuration item risk areas on technical, cost, and schedule basis, and (d) review the preliminary hardware product specifications. For a computer software configuration item (CSCI), this review will focus on the determination of the acceptability of the detailed design, performance, and test characteristics of the design solution, and on the adequacy of the operation and support documents.
Customer	The recipient or beneficiary of the outputs of the process work efforts, or the purchaser of its products and services. The customer may be either internal or external to the organization or company.
D Spec	Common name for the Process Specification as defined by MIL-STD-961. See further definition below.

Demonstration and Validation (Dem/Val) The government acquisition phase where one or more of the system concepts from the Concept Exploration and Definition phase is developed enough to demonstrate and validate that the concept is ready for Engineering and Manufacturing Development (EMD). Dem/Val is sometimes referred to as Phase 1.

This phase is now called "Program Definition and Risk Reduction" to emphasize a different focus for this phase.

Deployment Readiness Review (DRR) DRR is intended to determine the status of completion of specific activities required for deployment go-ahead decisions. DRR occurs incrementally during EMD, addressing all areas of concern in the Deployment Plan, with early stages devoted to gross level deployment concerns and progressing to a more detailed level as the design matures. DRR is complete when both contractor and government have verified that all activities required to support a deployment go-ahead decision have been completed.

Design-To-Cost (DTC) Management concept wherein rigorous cost goals are established during development. The control of system costs for the entire life cycle (development, manufacturing, verification, deployment, operation, support, training and disposal) to meet these DTC goals is achieved by practical trade-offs between operational capability, performance, costs, and schedule. Cost, as a key design parameter, is addressed on a continuing basis and as an inherent part of the development and production process.

Development Specification (B Spec) A document applicable to an item below the system level which states performance, interface and other technical requirements in sufficient detail to permit design, engineering for service use, and evaluation. Sometimes known as a Part I specification.

E Spec Common name for the Material Specification as defined by MIL-STD-961. See further definition below.

Engineering and Manufacturing Development (EMD) The government acquisition phase where the system design is fully developed in preparation for full scale production and deployment. Usually follows the Demonstration and Validation phase. EMD is sometimes referred to as Phase 2. Formerly known as Full Scale Development (FSD).

Exit Criteria Project accomplishments that must be satisfactorily demonstrated before an effort or project can progress further in the current development phase, or transition to the next development phase. Exit criteria may include such factors as critical test issues, the attainment of projected growth curves and baseline parameters, and the results of risk reduction efforts deemed critical for the decision to proceed further. Exit criteria supplement minimum required accomplishments and are specific to each development phase.

Formal Qualification Review (FQR) The objective of the FQR is to verify that the actual performance of the configuration items of the system, as determined through test, comply with the hardware Development Specification, Software Requirements and Interface Requirements Specifications, and to identify the test report(s)/data which document results of qualification tests of the configuration items. When feasible, the FQR is combined with the FCA at the end of configuration item/subsystem testing, prior to PCA. If sufficient test results are not available at the FCA to insure the configuration items will perform in their system environment, the FQR is conducted (post PCA) during System testing, whenever the necessary tests have been successfully completed to enable certification of configuration items.

Full Scale Development (FSD) Previous name for Engineering and Manufacturing Development (EMD) phase.

Function A task, action or activity that must be performed to achieve a desired outcome. A capability of the system or system element. May be static or dynamic.

Functional Analysis/ Allocation Examination of a defined function to identify all the subfunctions necessary to the accomplishment of that function. The subfunctions are arrayed in a functional architecture to show their relationships and interfaces (internal and external). Upper-level performance requirements are flowed down and allocated to lower-level subfunctions.

Functional Configuration Audit (FCA) A formal audit to validate that the development of a configuration item has been completed satisfactorily, and that the configuration item has achieved the performance and functional characteristics specified in the functional or allocated configuration identification. In addition, the completed operation and support documents shall be reviewed.

Inspection A formal or informal meeting of peers to inspect a document or other item for errors. Peer inspections should be performed in accordance with the In-Process Quality Inspection (IPQI) process document.

Integrated Logistics Support (ILS)	A disciplined, unified, and iterative approach to the management and technical activities necessary to integrate support considerations into system and equipment design; develop support requirements that are related consistently to readiness objectives, to design, and to each other; acquire the required support; and provide the required support during the operational phase at minimum cost.
Integrated Logistics Support (ILS) Elements	The principal elements of ILS include: Maintenance Planning; Manpower and Personnel; Supply Support; Support Equipment; Technical Data; Training and Training Support; Computer Resources Support; Facilities; Packaging, Handling, Storage and Transportation (PHS&T); and Design Interface.
Integrated Product Development (IPD)	Integrated Product Development (IPD) is a team activity to optimize the system within constraints involving engineering, manufacturing, test, configuration management, production support and business customers (the user, the maintainer, the trainer) that support the project manager. Achieving the desired results requires iterative product and process design changes using a structured process and teamwork among competent people with appropriate analysis tools and with specific expertise and understanding of the product, users, technology base, materials, manufacturing capabilities and requirements, training capabilities and requirements, support capabilities and requirements, and the acquisition process.
Integration	Act of putting together as the final end item various components of a system.

Also, refers to the integration of technical disciplines, as in "engineering integration." |
Integrator	The Integrator in government acquisition is the "prime prime" contractor.
Interface Control Document (ICD)	Interface Control Documents are used to define the functional, physical or operational interface between hardware, software, facilities, external systems and hardware/software items.
Interface Requirements Specification (IRS)	The Interface Requirements Specification describes in detail the requirements for one or more CSCI interfaces in the system, segment, or prime item.
Logistics Support	The supply and maintenance of materiel essential to proper operation of a system in the government force structure or in the customer premises.

**Logistics
Support
Analysis (LSA)**
The selective application of scientific and engineering efforts undertaken during the government acquisition process, as part of the SE process, to assist in: causing support considerations to influence design; defining support requirements that are related optimally to design and to each other; acquiring the required support; and providing required support during the operational phase at minimum cost.

**Logistics
Support
Analysis
Record
(LSAR)**
A formal tool under MIL-STD-1388-2A that uses records/forms to document operations and maintenance requirements, RAM (reliability, availability and maintainability), task analyses, technical data, support/test equipment, facilities, skill evaluation, supply support, ATE (automatic test equipment), TPS (test program/package set), and transportability. LSAR is the basis for training, personnel, supply provisioning and allowances construction, support equipment acquisition, facilities construction and preparation, and for preventative and corrective maintenance.

**Master Test
and Evaluation
Plan (MTEP)**
A document which identifies for each requirement how and when it will be verified, what level of verification, and the verification method. Requirements for special test equipment, software and facilities will be defined. Detailed test plans and test procedures will be developed based upon the MTEP. Usually the MTEP contains a Requirements Traceability Matrix (RTM).

**Material
Specification
(E Spec)**
A specification which defines the required qualities or condition of raw or semi-fabricated material used in fabrication.

**Measures of
Effectiveness
(MOEs)**
Metrics used to quantify the performance of system products and processes in terms that describe the utility or value when executing customer missions. Systems Engineering uses MOEs in a variety of ways including decision metrics, performance requirements, and in assessments of expected performance. MOEs can include cost effectiveness metrics.

**Measures of
Performance
(MOPs)**
Similar to MOEs except that MOPs are usually measurable with Technical Parameters (TPs) (such as bandwidth, time, beamwidth, etc.). MOEs can usually only be inferred from complicated models and are never really known until after the system is used in its real environment (such as a real war).

Mode
The condition of a system or subsystem in a certain state when specific capabilities (or functions) are valid. Typical modes in the Ready state, for example, are Normal, Emergency, Surge, Degraded, Reset, etc. Each mode may have different capabilities defined.

Operational Requirements Document (ORD)

A government document usually written by the actual or potential system users. It documents the users' objectives and minimum acceptable requirements for successful operational performance of a proposed concept or system. Format has been standardized across all DOD components by DOD Instruction 5000.1 and DOD 5000.2-M.

Non-government customers sometimes write a document of this type, but more often it would be written by a marketing organization to document the perceived needs of the customer.

The ORD is a type of User Requirements Document (URD).

Physical Configuration Audit (PCA)

The Physical Configuration Audit (PCA) shall be the formal examination of the as-built version of a configuration item against its design documentation, in order to establish the product baseline. After successful completion of the audit, all subsequent changes are processed by engineering change action. The PCA also determines that the acceptance testing requirements prescribed by the documentation are adequate for acceptance of production units of a configuration item by quality assurance activities. The PCA includes a detailed audit of engineering drawings, specifications, technical data and tests utilized in production of HWCIs, and a detailed audit of design documentation, listings, and manuals for CSCIs. The review shall include an audit of the released engineering documentation and quality control records to make sure the as-built or as-coded configuration is reflected by this documentation. For software, the Software Product Specification and Version Description Document shall be a part of the PCA review.

Preplanned Product Improvement (P3I)

Planned future evolutionary improvement of developmental systems for which design considerations are effected during development to facilitate future application of projected technology or features. Includes improvements planned for ongoing systems that go beyond the current performance envelope to achieve a needed operational capability.

Preliminary Design Review (PDR)
This review is conducted for each configuration item or aggregate of CIs to: (a) evaluate the progress, technical adequacy, and risk resolution of the selected design approach, (b) determine its compatibility with performance and engineering specialty requirements of the development specification, (c) evaluate the degree of definition and assess the technical risk associated with the selected manufacturing methods/processes, (d) establish the existence and compatibility of the physical and functional interfaces among the configuration item and other items of equipment, facilities, software, and personnel. Specifically, this review will focus on (a) evaluation of the progress, consistency, and technical adequacy of the selected top-level design and test approach, (b) compatibility between hardware/software requirements and preliminary design, and (c) on the preliminary version of the operation and support documents.

Primary Life Cycle Functions
The Primary Life Cycle Functions are those that must be considered when deriving the functional and performance requirements of a system:
1) Development
2) Manufacturing
3) Verification
4) Deployment
5) Operations
6) Support
7) Training
8) Disposition

Process Champion
The agent by which a process is introduced into a project and managed throughout the project's life. (See Appendix D for a summary of the Process Champion's roles and responsibilities in the SE process.)

Process Specification (D Spec)
A specification which defines the method by which resources are connected to usable products (through fabrication, solder, assembly, support, etc.).

Product Specification (C Spec)
A document applicable to a production item below the system level which states item characteristics in a manner suitable for procurement, production and acceptance. A product specification states (a) the complete performance requirements of the product for the intended use, and (b) the necessary interface and interchangeability characteristics. It covers form, fit and function. Production specifications are sometimes known as Part II specifications.

Production   PRR is intended to determine the status of completion of specific
Readiness   activities required for production go-ahead decisions. PRR occurs
Review (PRR)   incrementally during EMD, addressing all areas of concern in the
                        Manufacturing Plan, with early stages devoted to gross level
                        manufacturing concerns and progressing to a more detailed level as
                        the design matures. There is no upper limit to the number of
                        incremental PRRs; PRR schedules relate to major design milestones,
                        but are not specifically related to other Design Reviews.
                        Subcontractors/suppliers are included in PRRs. PRR is complete
                        when both contractor and Government have verified that all activities
                        required to support a production go-ahead decision have been
                        completed.

Project   Performing of the technical program management tasks that are not
Engineering   provided by the Project Management organization. Perform total
                        program management for those projects needing technically
                        knowledgeable personnel.

Qualification   The process of proving that a product or process meets all its
                        requirements. Design qualification verifies compliance to all A Spec
                        and B Spec requirements. Preproduction qualification verifies
                        compliance to all C Spec requirements. Predeployment qualification
                        verifies compliance to all D Spec requirements for deployment.

Quality   QFD is a technique for identifying customer needs and their relative
Function   priority. These needs are mapped to the required technical
Deployment   characteristics. Interaction between the technical characteristics is
(QFD)   identified and categorized as to whether there is a positive, neutral or
                        negative correlation. QFD is used to support trade studies and
                        marketing analysis. The QFD information is usually documented in a
                        "House of Quality" format.

Requirements   The determination of system specific characteristics based on analysis
Analysis   of customer needs, requirements and objectives; missions; projected
                        utilization environments for people, products and processes; and
                        measures of effectiveness. Requirements analysis assists the
                        customers in refining their requirements in concert with defining
                        functional and performance requirements for the system's primary life
                        cycle functions. It is a key link in establishing achievable
                        requirements that satisfy needs.

Requirements   Requirements Traceability Matrices show the allocation of
Traceability   requirements from the system specification to the CIs, other system
Matrix (RTM)   elements, functional areas, processes, external systems, etc. The
                        RTM also maps requirements to tests in the MTEP or discipline test
                        plans.

Segment Specification (A Spec)

A document which states the technical and mission requirements for a system segment as an entity, allocates requirements to functional areas (or configuration items), and defines the interfaces between or among the functional areas. A segment is a portion of a system and is usually higher than a configuration item.

Software Requirements Specification (SRS)

Software Requirements Specification describes in detail the functional, interface, quality factors, special, and qualification requirements necessary to design, develop, test, evaluate and deliver the required Computer Software Configuration Item (CSCI).

Software Specification Review (SSR)

A review of the finalized Computer Software Configuration Item (CSCI) requirements and the operational concept. The SSR is conducted when CSCI requirements have been sufficiently defined to evaluate the contractor's responsiveness to, and interpretation of, the system, segment, or prime item level requirements. A successful SSR is predicated upon the contracting agency's determination that the Software Requirements Specification (SRS), Interface Requirements Specification(s) (IRS), and Operational Concept Document (OCD) form a satisfactory basis for proceeding into preliminary software design.

Specification

A document intended primarily for use in procurement, which clearly and accurately describes the essential technical requirements for items, materials or services including the procedures by which it will be determined that the requirements have been met. The major types of specification documents (A, B, C, D, & E) are defined elsewhere in this Glossary.

Standard

Document that establishes engineering and technical requirements for processes, procedures, practices, and methods that have been decreed by authority or adopted by consensus. Standards may also be established for selection, application and design criteria for materiel.

State

The condition of a system or subsystem when specific modes or capabilities (or functions) are valid. Typical states are Off, Start-Up, Ready, On, Deployed, Stored, In-Flight, In-Service, etc.

Statement of Work (SOW)

That portion of a contract which establishes and defines all non-specification requirements for contractor efforts either directly or with the use of specific cited documents.

Subsystem

A grouping of items satisfying a logical group of functions within a particular system.

Synthesis   The translation of functions and requirements into possible solutions (resources and techniques) satisfying basic input requirements. System element alternatives that satisfy allocated performance requirements are generated. Preferred system element solutions that satisfy internal and external physical interfaces are selected. System concepts, preliminary designs and detailed designs are completed as a function of the development phase. System elements are integrated into a physical architecture.

System   A composite of subsystems, assemblies, skills, and techniques capable of performing and/or supporting an operational (or non-operational) role. A complete system includes related facilities, items, material, services, and personnel required for its operation to the degree that it can be considered a self-sufficient item in its intended environment.

System Analysis and Optimization   The process of assessing system effectiveness and risks, developing and validating system models, and performing trade studies. Must balance cost, schedule, performance and risk.

System Design Review (SDR)   This review shall be conducted to evaluate the optimization, correlation, completeness, and risks associated with the allocated technical requirements. Also included is a summary review of the systems engineering process which produced the allocated technical requirements and of the engineering planning for the next phase of effort. Basic manufacturing considerations will be reviewed and planning for production engineering in subsequent phases will be addressed. This review will be conducted when the system definition effort has proceeded to the point where system functional characteristics are defined and the configuration items are identified.

System Element   The basic constituents that comprise a system and satisfy one or more requirements in the lowest levels of the functional architecture.

System Requirements Review (SRR)   The objective of this review is to ascertain the adequacy of the contractor's efforts in defining system requirements. It will normally be conducted during the Concept Exploration or early in the Demonstration and Validation Phase when a significant portion of the system functional requirements has been established.

System Specification (A Spec)   A document which states the technical and mission requirements for a system as an entity, allocates requirements to functional areas (or configuration items), and defines the interfaces between or among the functional areas.

System Verification Review (SVR)	A review of the system verification results to ensure that all requirements in the A and B specs have been satisfied.
System/ Segment Design Document (SSDD)	A document which defines the system or segment design as it evolves. Includes descriptions of hardware, software, and manual operations (of personnel). Also includes system mission and operational concepts. Contains design implementations of B specs. This can be considered as a working document until the C specs have been fully developed. Normally only used on jobs where J-STD-016 is invoked, but the SSDD can be a useful document on other projects as well.
Systems Engineering (SE)	Systems Engineering is a comprehensive, iterative process that incorporates the technical efforts of the entire technical team to evolve and verify an integrated and optimally balanced set of product and process designs that satisfy user needs, and provides information for management decision making. Systems Engineering integrates the technical efforts of the technical team to meet project cost, schedule and performance objectives within an optimal design solution that encompasses the product and process aspects of an item. Systems Engineering derives a balanced product/process set by trading costs, schedule and risks with performance benefits of solution alternatives.
Systems Engineering Detailed Schedule (SEDS)	The detailed, task oriented schedule of the work efforts required to support the events and tasks identified in the SEMS.
Systems Engineering Management Plan (SEMP)	The Systems Engineering Management Plan (SEMP) is the primary, top-level technical management document for the integration of all systems engineering activities within the context of, and as an expansion to, the project plan. EIA 632 and IEEE 1220 define the contents of a SEMP.
Systems Engineering Manager (SEM)	The person responsible for managing the technical execution of a project. This task may be allocated to the Project Manager (if the project is small enough), the Deputy Project Manager, or to the Chief Engineer. The SEM must be responsible technically for the entire system, including any manufacturing and deployment development, if required. The SEM is not necessarily from a Systems Engineering organization, but must be skilled and experienced in the SE process.

Systems Engineering Master Schedule (SEMS)

The Systems Engineering Master Schedule (SEMS) documents the accomplishment criteria for all critical tasks and milestones in the SEMP and the Systems Engineering Detailed Schedule (SEDS). The SEMS uses metrics that include Technical Parameters (TPs) and specific success criteria for progress assessment.

The SEMS is supported by the Systems Engineering Detailed Schedule (SEDS). The SEDS identifies the detailed technical tasks required to accomplish each task/event in the SEMS. The SEDS is *time-based* while the SEMS is *event-based*.

Tailored SE Process

A variant of the generic systems engineering process which has been adapted to satisfy project-specific requirements.

Technical Parameter (TP)

A selected subset of the system's technical metrics tracked in TPM. Critical technical parameters are identified from risk analyses and contract specification or incentivization, and are designated by management. Examples of TPS include: a) Specification requirements; b) Metrics associated with technical objectives and other key decision metrics used to guide and control progressive development; c) Design-To-Cost (DTC) requirements; and d) Parameters identified in the acquisition program baseline or user requirements documentation.

Technical Performance Measurement (TPM)

The continuing verification of the degree of anticipated and actual achievement of technical parameters. TPM is used to identify and flag the importance of a design deficiency that might jeopardize meeting a system-level requirement that has been determined to be critical. Measured values that fall outside an established tolerance band require corrective actions to be taken by management.

Technical Reviews and Audits

Technical reviews are conducted to assess the degree of completion of technical efforts before proceeding beyond critical events and key project milestones. The schedule and plan for the conduct of technical reviews are included in the contractor's Systems Engineering Management Plan. Reviews are structured within the total system context to (a) demonstrate that the relationships, interactions, interdependencies and interfaces between required items (e.g., a communication interface), system functions, subsystems, configuration items and elements, as appropriate, have been addressed; (b) ensure that requirements are flowed down as required; and (c) ascertain the status/degree of completion of the product and process solution to those required.

Test and Evaluation Master Plan (TEMP)	A document used by a government Contracting Agency to document the overall test and evaluation plan, designed to identify and integrate objectives, responsibilities and schedules for all test and evaluation to be accomplished prior to subsequent key decision points. It is prepared by the government as early as possible in the acquisition process and is updated as development progresses. It is not always made available to contractors, but when it is, it is quite useful in helping to understand the customer's needs and requirements.
Test Readiness Review (TRR)	This review is conducted for each CSCI to determine whether the software test procedures are complete, and to assure readiness for formal CSCI testing. Software test procedures are evaluated for compliance with software test plans and descriptions, and for adequacy in accomplishing test requirements. The results of informal software testing and any updates to operational and support documents are also reviewed. A successful review will determine that the software test procedures and informal test results form a satisfactory basis for proceeding into formal CSCI testing.
Test Requirements Sheet (TRS) *or* **Test Requirements Specification (TRS)**	Test Requirements Sheet is a worksheet that identifies all the requirements that must be demonstrated or verified during the life cycle testing. A Test Requirements Specification is the formatted collection of all the worksheets. The TRS serves as a tool for management to check whether appropriate provisions have been made for verification of all performance/design requirements. It also provides for the identification of test functions for the test cycle of the systems engineering process.
Test, Analyze and Fix (TAAF)	Test, Analyze and Fix (TAAF) is a planned process in which development items are tested under actual or simulated mission profile environments to disclose design deficiencies, and to provide engineering information on failure modes and mechanisms. The purpose of TAAF is to provide a basis for early incorporation of corrective actions and verification of their effectiveness in improving the reliability of equipment.
User Requirements Document (URD)	A document that contains requirements of end users of the system or its products. Usually written by marketing, user research agencies, or some other organization. Sometimes this document is written by the users themselves. The requirements contained therein are often not stated in technical terms and need to be translated into specific domain terminology of the design engineering community by the SE process.

Work
Breakdown
Structure
(WBS)

A product-oriented family tree composed of hardware, software, data, facilities and services which result from systems engineering efforts during the development and production of system elements. Displays and defines the product(s) to be developed or produced, and relates the elements of work to be accomplished to each other and to the end product. Provides structure for guiding multi-disciplinary team assignment and cost tracking and control.

Acronyms

ABD	Architecture Block Diagram
AHP	Analytical Hierarchy Process
ANSI	American National Standards Institute
ASIC	Application Specific Integrated Circuit
ATR	Acceptance Test Review
CAE	Computer-Aided Engineering
CALS	Computer-Aided Acquisition Logistics Support, or Continuous Acquisition Life-Cycle Support
CAM	Computer-Aided Manufacturing
CAM	Cost Account Manager
CASE	Computer-Aided Software Engineering
CBS	Cost Breakdown Structure
CCB	Configuration Control Board
CDR	Critical Design Review
CDRL	Contract Data Requirements List
CDS	Concept Description Sheet
CE	Concurrent Engineering
CER	Cost Estimating Relationship
CI	Configuration Item
CIM	Computer Integrated Manufacturing
CM	Configuration Management

CO	Change Order
COEA	Cost and Operational Effectiveness Analysis
CPT	Cross-Project or Cross-Product Team
CR	Change Request
CSAR	Configuration Status Accounting Report
CSCI	Computer Software Configuration Item
CSE	Chief Systems Engineer
CSR	Change Status Report
CWBS	Contract Work Breakdown Structure
DCS	Design Constraint Sheet
DID	Data Item Description
DM	Data Management
DOD	Department of Defense
DRR	Deployment Readiness Review
DTC	Design-To-Cost
ECN	Engineering Change Notice
ECP	Engineering Change Proposal
EIA	Electronic Industries Association
EIMS	End Item Maintenance Sheet
EMD	Engineering and Manufacturing Development
FAT	Factory Acceptance Test (or First Article Test)
FCA	Functional Configuration Audit
FDD	Functional Design Discipline
FFBD	Functional Flow Block Diagram
FMA	Failure Mode Analysis
FMEA	Failure Mode and Effects Analysis
FMECA	Failure Mode, Effects and Criticality Analysis
FPGA	Floating Point Gate Array
FQR	Formal Qualification Review
FRACAS	Failure Reporting and Corrective Action System
FRB	Failure Review Board

FSD	Full Scale Development
HW	Hardware
HWCI	Hardware Configuration Item
I&V	Integration and Verification
ICD	Interface Control Document
ICWG	Interface Control Working Group
IDD	Interface Design Document
IEEE	Institute of Electrical and Electronics Engineers
ILS	Integrated Logistics Support
INCOSE	International Council on Systems Engineering
IPD	Integrated Product Development
IPPD	Integrated Process and Product Development
IPQI	In-Process Quality Inspection
IPT	Integrated Product Team
IR&D	Independent Research and Development
IRS	Interface Requirements Specification
ISO	International Standards Organization
IV&V	Independent Verification and Validation
KSA	Knowledge, Skills, and Abilities
KSAM	Knowledge, Skills, Abilities, and Motivations
LC	Life Cycle
LCC	Life-Cycle Cost
LSA	Logistics Support Analysis
LSAR	Logistics Support Analysis Record
MOE	Measure of Effectiveness
MOP	Measure of Performance
MR	Modification Request
MTEP	Master Test and Evaluation Plan
NASA	National Aeronautics and Space Administration
O&S	Operations and Support
OA&M	Operations, Administration and Maintenance

OBS	Organizational Breakdown Structure
OCD	Operational Concept Document
OO	Object Oriented
OOA	Object Oriented Analysis
OOD	Object Oriented Design
ORD	Operational Requirements Document
P3I	Preplanned Product Improvement
PC	Personal Computer
PCA	Physical Configuration Audit
PDCA	Plan, Do, Check, Act
PDR	Preliminary Design Review
PHS&T	Packaging, Handling, Storage and Transportation
PM	Project Manager
PMT	Process Management Team
PMTE	Process, Methods, Tools, and Environment
PRR	Production Readiness Review
PS	Production Sheet
QA	Quality Assurance
QFD	Quality Function Deployment
QIT	Quality Improvement Team
RAM	Responsibility Assignment Matrix
RAS	Requirements Allocation Sheet
RFP	Request for Proposal
RRA	Roles, Responsibility, and Authority
RTM	Requirements Traceability Matrix
RVM	Requirements Verification Matrix
RVTM	Requirements Verification Traceability Matrix
SAT	Site Acceptance Test
SBD	Schematic Block Diagram
SBS	System Breakdown Structure
SDD	Software Design Document

SDP	Software Development Plan
SDR	System Design Review (being replaced by SFR)
SDRL	Subcontract Data Requirements List
SE	Systems Engineering
SEDE	Systems Engineering Development Environment
SEDS	Systems Engineering Detailed Schedule
SEM	Systems Engineering Manager
SEMP	Systems Engineering Management Plan
SEMS	Systems Engineering Master Schedule
SI&V	System Integration and Verification
SIWG	System Integration Working Group
SOW	Statement of Work
SpE	Specialty Engineering
SPS	Software Product Specification
SRR	System Requirements Review
SRS	Software Requirements Specification
SSDD	System/Segment Design Document
SSR	Software Specification Review
SVR	System Verification Review
SW	Software
T&E	Test and Evaluation
TAAF	Test, Analyze and Fix
TEMP	Test and Evaluation Master Plan
TLS	Time Line Sheet
TP	Technical Parameter
TPM	Technical Performance Measurement
TQM	Total Quality Management
TRR	Test Readiness Review
TRS	Test Requirements Specification (or Sheet)
URD	User Requirements Document
WBS	Work Breakdown Structure

Bibliography

"Books are the quietest and most constant of friends;
they are the most accessible and wisest of counselors,
and the most patient of teachers."

—Charles W. Eliot

"Books are good enough in their own way, but they are
a might bloodless substitute for life."

—Robert Louis Stevenson

A. General Systems Engineering

1. Aslaksen, Erik and Rod Belcher, *Systems Engineering.* Prentice Hall, 1992.
2. Beam, Walter R., *Systems Engineering: Architecture and Design.* McGraw Hill, 1990.
3. Belev, G. C., "Guidelines for Specification Development," *Proceedings of the Annual IEEE Symposium on Reliability and Maintainability.* 1989.
4. Blanchard, B., and W. Fabrycky, *Systems Engineering and Analysis,* 2nd ed. Prentice Hall, 1990.
5. Boardman, John, *Systems Engineering: An Introduction.* Prentice Hall, 1990.
6. Chestnut, H., *System Engineering Tools.* John Wiley, 1965.
7. Chestnut, H., *Systems Engineering Methods.* John Wiley, 1967.
8. Dandy, G. C. and R. F. Warner, *Planning and Design of Engineering Systems.* Unwin Hyman, 1989.

9. Drew, D. R., and C. H. Hsieh, *A Systems View of Development: Methodology of Systems Engineering and Management.* Cheng Yang Publishing Co., No. 4, Lane 20, Gong-Yuan Road, Taipei, ROC, 1984.

10. Electronic Industries Association, *System Engineering.* SYSB-1, EIA Engineering Bulletin, December 1989.

11. Gheorge, A., *Applied Systems Engineering.* UMI, 1994 reprint (originally published by Wiley, 1984).

12. Grady, J. O., *System Requirements Analysis.* McGraw Hill, 1993.

13. Hoban, F. T. and W. M. Lawbaugh, eds. *Readings in Systems Engineering.* NASA, 1993.

14. Hunger, Jack W., *Engineering the System Solution: A Practical Guide to Developing Systems.* Prentice Hall, 1995.

15. Machol, R. E., ed., *System Engineering Handbook.* McGraw Hill, 1965.

16. Mar, Brian, *Introduction to the Engineering of Complex Systems.* Class Notes, University of Washington, 1994.

17. Meredith, D. D., et al, *Design and Planning of Engineering Systems.* Prentice Hall, 1973.

18. Ostrofsky, B., *Design, Planning and Development Methodology.* Prentice Hall, 1977.

19. Purdy, D. C., *A Guide to Writing Successful Engineering Specifications.* McGraw Hill, 1991.

20. Reilly, Norman B., *Successful Systems for Engineers and Managers.* Van Nostrand Reinhold, 1993.

21. Sage, A. P., *Methodology for Large Scale Systems.* McGraw Hill, 1977.

22. Sage, A. P., *Systems Engineering.* Wiley, 1992.

B. Special Systems Engineering

1. Beam, W. R., *Command, Control, and Communications Systems Engineering.* McGraw Hill, 1989.

2. Crinnion, John, *Evolutionary Systems Development.* Plenum, 1991.

3. Gall, John, *Systemantics: The Underground Text of Systems Lore, How Systems Really Work and How They Fail.* General Systemantics Press, 1986.

4. Gershwin, Stanley B., *Manufacturing Systems Engineering.* Prentice Hall, 1994.

5. Quintas, Paul, ed., *Social Dimensions of Systems Engineering.* Ellis Horwood, 1993.

6. Sage, A. P., *Decision Support Systems Engineering.* Wiley, 1991.

7. Sage, Andrew P. and James D. Palmer, *Software Systems Engineering.* Wiley, 1990.

8. Saracco, R., et al, *Telecommunications Systems Engineering Using SDL*. North Holland, 1989.
9. Schiebe, Michael, et al., eds., *Real-Time Systems Engineering and Applications*. Kluwer Academic, 1992.
10. Thome, Bernhard, ed., *Systems Engineering: Principles and Practice of Computer-based Systems Engineering*. Wiley, 1993.
11. Wymore, A. Wayne, *A Mathematical Theory of Systems Engineering: The Elements*. Krieger, 1977 reprint (original Wiley, 1967).
12. Wymore, A. Wayne, *Model-Based Systems Engineering*. CRC Press, 1993.

C. Systems Architecting and Architecture

1. Alexander, Christopher, *Notes on the Synthesis of Form*. Harvard, 1964.
2. Chorafas, Dimitris N., *Systems Architecture and Systems Design*. McGraw Hill, 1989.
3. Rechtin, E. and Mark Maier, *The Art of Systems Architecting*. CRC Press, 1996.
4. Rechtin, E., *Systems Architecting: Creating & Building Complex Systems*. Prentice Hall, 1991.

D. Reference Books and Handbooks

1. AT&T, *Design Policy, Design Process, Design Analysis*. Reference Guide. AT&T, Draft, December 1989.
2. AT&T, *Design Reference Mission Profile, Design Requirements, Trade Studies*. Reference Guide. AT&T, Draft. December 1989.
3. AT&T, *Design Reviews*. Reference Guide. AT&T, Draft. December 1989.
4. AT&T, *Front End Process*. Reference Guide. AT&T, Draft. December 1989.
5. AT&T, *Systems Engineering Process*, 2nd ed. AT&T ATS Engineering Standard Process, November 1994.
6. Dorfman, M., and R. H. Thayer, eds., *Standards, Guidelines, and Examples on System and Software Requirements Engineering*. IEEE Computer Society Press, 1990.
7. Thayer, R. H., and M. Dorfman, eds., *System and Software Requirements Engineering*. IEEE Computer Society Press, 1990.

E. General Engineering, Concurrent Engineering, and Design Engineering

1. Beakley, G. C., D. L. Evans, and J. B. Keats, *Engineering – An Introduction to a Creative Profession*, 3rd ed. Macmillan, 1986.

2. Dieter, G. E., *Engineering Design: A Materials and Processing Approach.* McGraw Hill, 1983.

3. Harper, C. A., ed., *Handbook of Electronic Systems Design.* McGraw Hill, 1980.

4. Petroski, Henry, *Design Paradigms: Case Histories of Error and Judgment in Engineering.* Cambridge, 1994.

5. Petroski, Henry, *The Evolution of Useful Things.* Knopf, 1992.

6. Pugh, S., *Total Design: Integrated Methods for Successful Product Engineering.* Addison-Wesley, 1991.

7. Shina, S. G., *Concurrent Engineering and Design for Manufacture of Electronic Products.* Van Nostrand Reinhold, 1991.

8. Shina, S. G., ed., *Successful Implementation of Concurrent Engineering Products and Processes.* Van Nostrand Reinhold, 1994.

9. Woodson, T. T., *Introduction to Engineering Design.* McGraw Hill, 1966.

F. Systems Engineering Management

1. Augustine, N. R., *Augustine's Laws and Major System Development Programs.* American Institute of Aeronautics and Astronautics, 1983.

2. Blanchard, B. S., *System Engineering Management.* John Wiley, 1991.

3. Chase, W. P., *Management of System Engineering.* John Wiley, 1974.

4. Grady, Jeffrey O., *System Engineering Planning and Enterprise Identity.* CRC Press, 1995.

5. Lacy, J. A., *Systems Engineering Management: Achieving Total Quality.* McGraw Hill, 1992.

G. General Engineering Management

1. Baumgartner, J., *Systems Management.* UMI, 1994 reprint (originally published by Bureau of National Affairs, 1979).

2. Blanchard, B. S., *Engineering Organization and Management.* Prentice Hall, 1976.

3. Blanchard, Frederick L., *Engineering Project Management.* Marcel Dekker, 1990.

4. Bockrath, J. T., *Dunham and Young's Contract, Specifications, and Law for Engineers*, 4th ed. McGraw Hill, 1986.

5. Clark, Kim B. and S. C. Wheelwright, *Managing New Product and Process Development: Text and Cases.* Harvard, 1993.

6. Cleland, D. I. and H. Kerzner, *Engineering Team Management.* Krieger, 1986.

7. Coutinho, John de S., *Advanced Systems Development Management.* Krieger, 1984.

8. Dhillon, B. S., *Engineering Management: Concepts, Procedures and Models*. Technomic, 1987.
9. Fleming, Q. W., *Cost/Schedule Control Systems Criteria: The Management Guide to C/SCSC*. Probus Publishing, 1988.
10. Helgeson, Donald V., *Engineer's and Manager's Guide to Winning Proposals*. Artech, 1994.
11. Jackson, M. C., *Systems Methodology for the Management Sciences*. Plenum Press, 1991.
12. Johnson, R. A., et al, *The Theory and Management of Systems*, 3rd ed. McGraw Hill, 1973.
13. Karger, D. W., and R. G. Murdick, *Managing Engineering and Research*. Industrial Press, 1969.
14. Michaels, J. V., and W. P. Wood, *Design to Cost*. John Wiley, 1989.
15. O'Connor, P. D. T., *The Practice of Engineering Management: A New Approach*. Wiley, 1994.
16. Peterson, Robert O., *Managing the Systems Development Function*. Van Nostrand Reinhold, 1987.
17. Turtle, Q. C., *Implementing Concurrent Project Management*. Prentice Hall, 1994.

H. Integration and Test

1. AT&T, *Testing to Verify Design and Manufacturing Readiness*. McGraw Hill, 1993.
2. Grady, Jeffrey O., *System Integration*. CRC Press, 1994.

I. Modeling and Simulation

1. Chapman, W. L., et al., *Engineering Modeling and Design*. CRC Press, 1992.
2. Ingels, D. M., *What Every Engineer Should Know about Computer Modeling and Simulation*. Marcel Dekker, 1985.
3. Rivett, P., *Principles of Model Building*. John Wiley, 1972.

J. Systems Theory, System Science and Problem Solving

1. Ackoff, R. L., *The Art of Problem Solving*. John Wiley, 1978.
2. Checkland, Peter, *Systems Thinking, Systems Practice*. Wiley, 1981.
3. De Bono, E., *de Bono's Thinking Course*. Facts on File Publications, 1985.
4. Forrester, Jay W., *Principles of Systems*. Productivity Press, 1971.
5. Hall, A. D., *A Methodology for Systems Engineering*. Van Nostrand, 1962.

6. Hall, A. D., *Metasystems Methodology: A New Synthesis and Unification.* Pergamon Press, 1989.

7. Jackson, M. C. and P. Keys, eds., *Systems Thinking in Action.* Journal of the Operational Research Society, Vol. 36, No. 9, September 1985.

8. Jackson, M. C. and Robert L. Flood, *Creative Problem Solving: Total Systems Intervention.* Wiley, 1991.

9. Luce, R. D., and H. Raiffa, *Games and Decisions.* John Wiley, 1957.

10. Miels, R., ed., *Systems Concepts.* John Wiley, 1973.

11. Optner, S. L., *Systems Analysis for Business and Industrial Problem Solving.* Prentice Hall, 1965.

12. Rubinstein, M. F., *Tools for Thinking and Problem Solving.* Prentice Hall, 1986.

13. Sandquist, G. M., *Introduction to System Science.* Prentice Hall, 1985.

14. Singh, M. G., ed., *Systems and Control Encyclopedia: Theory, Technology, Applications.* Pergamon Press, 1989.

15. Von Bertalanffy, L., *General Systems Theory.* George Braziller, 1968.

16. Warfield, John N., *Societal Systems.* Intersystems Publications, 1989.

17. Weinberg, Gerald, *An Introduction to General Systems Thinking.* Wiley, 1975.

18. Weinberg, Gerald, *Rethinking Systems Analysis & Design.* Dorset House, 1988.

19. Weiner, N., *Cybernetics.* John Wiley, 1948.

K. Creativity, Critical Thinking, Information Theory and Information Science

1. Adams, James L., *Conceptual Blockbusting: A Guide to Better Ideas.* W. W. Norton, 1979.

2. Bailey, Robert L., *Disciplined Creativity.* Ann Arbor Science, 1978.

3. Dick, Michael J., *"High Tech" Creativity.* IEEE, 1992.

4. Ruchlis, Hy, *Clear Thinking: A Practical Introduction.* Prometheus, 1990.

5. Ruggiero, Vincent Ryan, *The Art of Thinking: A Guide to Critical and Creative Thought.* Harper & Row, 1984.

6. Tufte, Edward R., *Envisioning Information.* Graphics Press, 1990.

7. Tufte, Edward R., *The Visual Display of Quantitative Information.* Graphics Press, 1983.

8. Volk, Tyler, *Metapatterns: Across Space, Time, and Mind.* Columbia, 1995.

L. Decision Making , Decision Support, and Risk Management

1. Andriole, Stephen J., *Handbook of Decision Support Systems*. TAP Books, 1989.
2. AT&T, *Design to Reduce Technical Risk*. McGraw Hill, 1993.
3. AT&T, *Moving a Design into Production*. McGraw Hill, 1993.
4. Fang, L., et al., *Interactive Decision Making: The Graph Model for Conflict Resolution*. Wiley, 1993.
5. Goldratt, E. M., *(What is this thing called) Theory of Constraints (and how should it be implemented?)*. North River Press, 1990.
6. Goldratt, E. M., *The Goal*. North River Press, 1992.
7. Morgan, M. G. and M. Henrion, *Uncertainty: A Guide to Dealing with Uncertainty in Quantitative Risk and Policy Analysis*. Cambridge, 1990.

M. Economic Analysis, Operations Research, Optimization, Trade Studies

1. Banks, J., and W. J. Fabrycky, *Procurement and Inventory Systems Analysis*. Prentice Hall, 1987.
2. Buffa, E. S., *Modern Production and Operations Management*, 5th ed. John Wiley, 1987.
3. Chase, R. B., and N. J. Aquilano, *Production and Operations Management: A Life Cycle Approach*. Richard D. Irwin, 1981.
4. Churchman, C. W., R. L. Ackoff, et al, *Introduction to Operations Research*, John Wiley, 1957.
5. Erikson, R. A., *Measures of Effectiveness in Systems Analysis and Human Factors*. Naval Weapons Center, China Lake, CA, September 1986.
6. Fabrycky, W. J., and G. J. Thuesen, *Economic Decision Analysis, 2nd ed.* Prentice Hall, 1980.
7. Fabrycky, W. J., P. M. Ghare, and P. E. Torgersen, *Applied Operations Research and Management Science*. Prentice Hall, 1984.
8. Hillier, F. S., and G. J. Lieberman, *Operations Research*, 4th ed. Holden-Day, 1986.
9. Pike, R. W., *Optimization for Engineering Systems*. Van Nostrand Reinhold, 1986.
10. Sage, A. P., *Economic System Analysis: Microeconomics for Systems Engineering, Engineering Management, and Project Selection*. Elsevier Science Publishing Co., 1983.
11. Schmid, H., et al, "CATS: Computer-Aided Trade Study Methodology," *IEEE National Aerospace and Electronics Conference*, 1987.

12. Stradler, W., *Multicriteria Optimization in Engineering and in the Sciences*. Plenum Press, 1988.

N. CAE, CAD, CAM, CALS

1. Eisner, H., *Computer-Aided Systems Engineering*. Prentice Hall, 1988.
2. Krouse, J. K., *What Every Engineer Should Know About Computer-Aided Design and Computer-Aided Manufacturing*. Marcel Dekker, 1982.
3. Teicholz, E., ed., *CAD/CAM Handbook*. McGraw Hill, 1985.

O. Software Engineering[54]

1. Brooks, Frederick P., *The Mythical Man-Month*. Addison Wesley, 1975.
2. Davis, A. M., *Software Requirements: Analysis and Specification*. Prentice Hall, 1990.
3. DeMarco, T., *Controlling Software Projects*. Yourdon Press, 1982.
4. DeMarco, T., *Structured Analysis and System Specification*. Prentice Hall, 1979.
5. Fairley, R. M., *Software Engineering Concepts*. McGraw Hill, 1985.
6. Fisher, A. S., *Using Software Development Tools*. Wiley, 1988.
7. Gehani, N., and A. D. McGettrick, eds., *Software Specification Techniques*. Addison-Wesley, 1986.
8. Hatley, D., and I. Pirbhai, *Strategies for Real-Time System Specification*. Dorset House Publishing, 1988.
9. Jirotka, M. and J. Goguen, *Requirements Engineering: Social and Technical Issues*. Academic, 1994.
10. Martin, J., and C. McClure, *Structured Techniques: The Basis for CASE*. Revised ed. Prentice Hall, 1988.
11. Pressman, R. S., *Software Engineering: A Practitioner's Approach*. McGraw Hill, 1982.
12. Sarson, C., *Structured Systems Analysis: Tools and Techniques*. Prentice Hall, 1979.
13. Vick, C. R., and C. V. Ramamoorthy, eds., *Handbook of Software Engineering*. Van Nostrand Reinhold, 1984.
14. Ward, P., and S. Mellow, *Structured Development for Real-Time Systems*. Prentice Hall, 1985.
15. Yourdon, E., *Managing the Structured Techniques*. Prentice Hall, 4th edition, 1989.
16. Yourdon, E., *Modern Structured Analysis*. Prentice Hall, 1989.

---

[54]For object oriented methods, see separate category in the Bibliography.

P. Object Oriented Analysis and Design

1. Coad, P., and E. Yourdon, *Object Oriented Analysis*. Prentice Hall, 1990.
2. Martin, James and James J. Odell, *Object-Oriented Analysis and Design*. Prentice Hall, 1992.
3. Martin, James, *Principles of Object-Oriented Analysis and Design*. Prentice Hall, 1993.
4. Rumbaugh, James, et al., *Object-Oriented Modeling and Design*. Prentice Hall, 1991.
5. Shlaer, S., and S. Mellor, *Object-Oriented System Analysis*. Prentice Hall, 1988.
6. Ward, P., *How to Integrate Object Orientation with Structured Analysis and Design*. IEEE Software, 1989.
7. Wegner, P., "Object Oriented Software Engineering," *ACM Tutorial Notes*. ACM, 1988.

Q. Defense Documents[55]

1. AR 70-1, Department of the Army, *Systems Acquisition Policy and Procedures*. 10 November 1988.
2. Defense Systems Management College, *Acquisition Strategy Guide*. 1984.
3. Defense Systems Management College, *Systems Engineering Management Guide*, 3rd ed. Fort Belvoir, Virginia, January 1990.
4. Department of Defense, *Total Quality Management Master Plan*. 1986.
5. DI-E-1135, Data Item Description, *Telecommunications System Engineering Plan*. Department of Defense, 2 May 1977.
6. DI-S-3618/S-152, Data Item Description, *System Engineering Management Plan (SEMP)*. Department of Defense, 9 February 1970.
7. DOD 4245.7-M, Department of Defense, *Transition from Development to Production*. September 1985.
8. DOD-STD-2167A, Military Standard, *Defense System Software Development*. Department of Defense, 29 February 1988.
9. DOD-STD-2168, Military Standard, *Defense System Software Quality Program*. Department of Defense, 25 April 1988.
10. FM 770-78, Department of the Army, Field Manual, *System Engineering*. April 1979.

---

[55]Even though many of these military standards have been canceled due to acquisition reform, they are still useful as guidance in implementing the practices and underlying principles contained therein.

11. MIL-HDBK-61, Military Handbook, *Configuration Management Guidance.* (to be published in Fall 1996).

12. MIL-S-83490, Military Specification, *Specifications, Types and Forms.* Department of Defense, 30 October 1968.

13. MIL-STD-1521B, Military Standard, *Technical Reviews and Audits for Systems, Equipments, and Computer Software.* Department of Defense, 4 June 1985, plus Notice 1, 19 December 1985.

14. MIL-STD-480B, Military Standard, *Configuration Control – Engineering Changes, Deviations and Waivers.* Department of Defense, 15 July 1988.

15. MIL-STD-483A, Military Standard, *Configuration Management Practices for Systems, Equipment, Munitions and Computer Programs.* Department of Defense, 4 June 1985.

16. MIL-STD-490A, Military Standard, *Specification Practices.* Department of Defense, 4 June 1985.

17. MIL-STD-498, Military Standard, *Software Development.* Department of Defense, 1995 (will be replaced by J-Std-016 and US 12207).

18. MIL-STD-499A, Military Standard, *Engineering Management.* Department of Defense, 1 May 1974.

19. MIL-STD-499B, Military Standard, *Systems Engineering.* Department of Defense, DRAFT, 6 May 1992.

20. MIL-STD-973, Military Standard, *Configuration Management.* Department of Defense, July 1992.

21. NAVSO P-6071, Department of the Navy, *Best Practices: How to Avoid Surprises in the World's Most Complicated Technical Process; The Transition from Development to Production.* March 1986.

22. UDI-E-23974, Data Item Description, *Plan, Systems Engineering Management (SEMP).* Department of Defense, 10 November 1972.

R. Specialty Engineering, Integrated Logistics Support (ILS) & Configuration Management (CM)

1. Blanchard, B. S., *Logistics Engineering and Management.* Prentice Hall, 1986.

2. Eggerman, W. V., *Configuration Management Handbook.* TAB Books, 1990.

3. Fabrycky, W. J., "Designing for the Life Cycle," *Mechanical Engineering.* January 1987.

4. Green, L. L., *Logistics Engineering.* John Wiley, 1991.

5. Jones, J. V., *Integrated Logistic Support Handbook.* TAB Books, 1987.

6. Jones, J. V., *Logistic Support Analysis Handbook.* TAB Books, 1989.

7. Woodson, W. E., *Human Factors Design Handbook.* McGraw Hill, 1981.

S. Quality Engineering and Needs Analysis

1. AT&T, *Process Quality Management & Improvement Guidelines*, Issue 1.1. AT&T Quality Steering Committee, 1988.
2. Clausing, Don, *Total Quality Development*. ASME Press, 1994.
3. Cohen, Lou, *QFD: How to Make QFD Work for You*. Addison-Wesley, 1995.
4. DOD 4245.7-M, *Transition from Development to Production*. Department of Defense, September 1985.
5. Duncan, A. J., *Quality Control and Industrial Statistics*, 4th ed. Richard D. Irwin, 1974.
6. Juran, J. M., and F. M. Gryna, eds., *Juran's Quality Control Handbook*, *4th ed.* McGraw Hill, 1988.
7. Juran, J. M., *Juran on Quality Improvement: Workbook*. Juran Enterprises, 1982.
8. NAVSO P-6071, *Best Practices: How to Avoid Surprises in the World's Most Complicated Technical Process*. Department of the Navy, March 1986.
9. Phadke, M. S., *Quality Engineering Using Robust Design*. Prentice Hall, 1989.
10. Priest, J. W., *Engineering Design for Producibility and Reliability*. Marcel Dekker, 1988.
11. *Quality Function Deployment for Product Definition*, AT&T Bell Laboratories Quality Technology and Services Center, May 1990.
12. Taguchi, S., *Taguchi Methods: Quality Engineering*. American Supplier Institute Press, 1988.

T. Industry Standards

1. EIA/IS 632, *Systems Engineering*. EIA, December 1994.
2. EIA/IS 649, *Configuration Management*. EIA, August 1995.
3. IEEE 1220, *Standard for Application and Management of the Systems Engineering Process*. IEEE, December 1994.
4. ISO/IEC 12207, *Software Life Cycle Processes*. ISO/IEC, February 1995.
5. J-Std-016, *Software Development*. EIA/IEEE, (to be published in late 1996).

appendix A

Definition of systems engineering

This Appendix is a reprint of two papers presented at the 1996 INCOSE Symposium. The first paper is by Dr. Jerry Lake and does an excellent job of defining "what systems engineering is" and "what systems engineering does." The second paper is by Richard Harwell and explains why and how systems engineering is "more than just a process." Both are used with permission.

Unraveling the Systems Engineering Lexicon

Dr. Jerome (Jerry) G. Lake
Systems Management international
281 So. Pickett St. #401
Alexandria, VA 22304

Abstract. Terms related to systems engineering have had a tempestuous history, at least for the years that the International Council on Systems Engineering (INCOSE) has been in existence. The term "Systems Engineering" has been at the center of the confusion.

This paper looks at the history and problems associated with defining systems engineering and then applies a systems engineering approach to provide definitions of what systems engineering is and what systems engineering does.

The definitions provided in this paper are not to end the discussion but to stimulate more discussion.

INTRODUCTION

Systems engineering is not a new concept. Several papers [e.g., Woods 1993; Alessi et al. 1995] and books [e.g., Chase 1974; Goode and Machol 1959] have referenced the concept back to at least the years of World War II. Systems engineering within enterprises, especially those doing work with the Department of Defense, has had ebbs and flows of importance during the period since World War II. The history of systems engineering is well documented and will not be rehearsed [sic] here. The fact that after so many years there is controversy about the role of systems engineering, about what it is and about what it does deserves the attention given over recent years and the attention given in this paper.

Early system engineers were not graduates of systems engineering oriented academic programs but were mathematicians, electrical and aerospace engineers, and other scientific disciplines. Common attributes of these early systems engineering pioneers included their tendency to be "big picture" thinkers, their capacity to accomplish integration of multidisciplinary inputs; their capability to solve complex problems; and their ability to conduct trade studies, effectiveness assessments, operations research, and requirements analyses.

Because of the importance of the concept, academic units created bachelors and graduate degree programs to prepare engineers with the skills and abilities to conduct systems engineering activities. Additionally, enterprises formed systems engineering organizations with individuals with the title of system engineer. And, noted holistic thinkers wrote books trying to capture the concept and what needs to be accomplished to solve the ever increasing complex problems facing the engineering community. One of the objectives of the International Council on Systems Engineering (INCOSE)[56] is to nurture academic programs to create engineers who can think in terms of the total system and who have the attributes of the early systems engineering pioneers.

Studies by INCOSE academics (e.g., Holtzman, Mar, Unwin) during the period 1991 through 1993 identified the several academic institutions offering programs in systems engineering. The researchers found that many degree programs that offer courses related to systems engineering do not carry the title "Systems Engineering." The lack of uniformity of program titles has been in part for the purpose of market discrimination of individual programs, but also because recognition or understanding of the term systems engineering is lacking.

It is the purpose of this paper to discuss why there appears to be so little agreement on the value of systems engineering or acceptance of the concept,

---

[56] Initially the Council was known as the National Council on Systems Engineering (NCOSE). It became known as INCOSE in July of 1995.

to look at the impact of lack of acceptance, and to remove some of the confusion as to what systems engineering was intended to be and what its function or purpose is. Definitions as to what systems engineering is and what it does are provided, not to end the discussion but to stimulate more discussion.

HISTORY

In August of 1990 a group of 35 individuals, recognized for their interest, knowledge, and expertise in systems engineering, met in Seattle to discuss the need for a professional organization to foster the definition and understanding of systems engineering. The cost of attendance was submission of the individual's concept or definition of systems engineering. The submitted definitions were categorized into four groups. The first group believed systems engineering to be a functional discipline of system engineers with the practice restricted to technical activities. The second group considered systems engineering to be a discipline of system engineers with practice including both technical and management activities. The third group expressed systems engineering as a set of technical activities practiced by any required discipline, not just system engineers. And, the fourth group believed systems engineering to be a set of technical and management activities practiced by any required discipline.

Thus, from the beginning, INCOSE has been composed of persons with differing views of what systems engineering is. It is because of these different views that INCOSE did not attempt to define the term. Doing so was considered as being counter productive to the purpose and objectives of the new Council. The intent of the founders was to include all who shared the vision that systems could be developed and produced at a lower cost, delivered on schedule, and provided with expected performance attributes if more system qualified engineers and better processes, methods and automated tools were available.

Recognition of Systems Engineering

In the years since inception of INCOSE, enterprises have been found in which system engineers, systems engineering organizations, and the term systems engineering itself are not recognized as needed disciplines, organizations, or practices. To such enterprises, systems engineering is analogous to high overhead, too much unnecessary documentation, and other costly Department of Defense oversight practices. This is not to say that these enterprises do not accomplish systems engineering. They do, and in many enterprises with relative success. They could, however, like most enterprise with system engineers, systems engineering organizations and established systems engineering practices, accomplish the engineering of systems more efficiently and effectively if a disciplined, comprehensive systems engineering approach or process were utilized.

Enterprises that have shunned systems engineering because of perceived negative connotations, have adopted and fostered terms such as Concurrent

Engineering and Integrated Product (and Process) Development. The trend lately is for enterprises with a history of embracing systems engineering, including the Department of Defense, to adopt these new terms to avoid the confusion related to systems engineering. Much of this can be directly attributed to a lack of appreciation of exactly what systems engineering is.

Commercial Systems Engineering Standards

This confusion and the perceived need for other terms to describe the engineering of systems is unfortunate, but understandable. The animosity (often hostility) toward systems engineering is one reason new commercial systems engineering standards (EIA/IS 632 and IEEE 1220-1994) have not received wide acceptance, even within INCOSE. Whereas system engineers look at these standards as much broader than what they do, non system engineers look at the standard titles and come up with the conclusion that the standards are not relevant to their work; that they are only applicable to system engineers. The adage "don't judge a book by its cover" is applicable in this case.

The intent of these two new standards is well defined in their respective forward and scope sections. The purpose of the EIA standard is to improve the engineering of systems in oversight or contractual instances. The IEEE standard purpose "is to provide a standard for managing a system from initial concept through development, operations, and disposal." It explains what any enterprise must do to engineer a system. Its focus is on commercial, non-oversight developments.

Neither standard restricts its tasks to what a system engineer does or for which a systems engineering organization is responsible. Specifically, teams and teamwork are called out to include personnel to ensure that quality factors related to producibility, testability/verifiability, deployability, operability, supportability, trainability, and disposability are designed into system products. Additionally, both standards call for inclusion, as appropriate, of customers/users, subcontractors, and other non-engineering personnel such as marketing, legal and contracting on interdisciplinary teams.

Two processes important to engineering a system are the central focus of the IEEE systems engineering standard. These are the life-cycle development process and the systems engineering process. The systems engineering process is the engine, recursively applied, that drives the evolution and maturity of the system through successive stages of development.

Systems Engineering in Relation to Concurrent Engineering and Integrated Process and Product Development

When the systems engineering envisioned in the standards is compared with the explained concepts and scope of Concurrent Engineering and Integrated Product and Process Development one finds that the purpose and scope of life cycles tasks are essentially the same, that the focus on downstream specialties in upstream design activities is the same, and that the utilization of automated tools is the same. The essential difference, which could lead one to consider the systems engineering approach described in the IEEE and EIA standards as being more comprehensive, is the inclusion of a comprehensive systems engineering process utilized recursively to mature the system solutions, attain customer satisfaction, and satisfy organizational commitments and public expectations.

So, the problems associated with acceptance of systems engineering appear to be based on lack of understanding on just what systems engineering is and what its purpose is.

One Definition of Systems Engineering

In July 1995, I found the following definition for systems engineering on the INCOSE Home Page on the World Wide Web. This definition was provided on the web in response to the question "What is Systems Engineering?" The author of the definition was not given. This definition includes not only what systems engineering is but also what it does.

Systems engineering is the discipline of managing the development of complex systems. It focuses on defining required functionality early in the development cycle, documenting these requirements, then proceeding with design synthesis while considering the complex problem: performance, manufacturing, cost and schedule, quality, training, and disposal. Systems engineering integrates all disciplines and specialty groups under one umbrella, employing a structured design process that facilitates the transition from concept to production to operation in an orderly fashion. Systems engineering considers both the business and technical needs of all customers - both users and suppliers.

Some, including me, may argue with some of the terms and phrases of this definition. For instance, this definition calls systems engineering a discipline. If by discipline the author means that accomplished by a "system engineer," then the definition would be unacceptable in that systems engineering is not just what a system engineer does. Systems engineering is not just another branch of engineering. Systems engineering encompasses work accomplished by any engineer or non-engineer who accomplishes tasks or activities related to engineering a system. However, if the author of the definition means that the practice of systems engineering is disciplined in that it includes a mechanism that can be used to ensure that a process is adequately followed, then the term disciplined is acceptable. There are other concepts expressed in this definition, such as *the discipline of managing* and *considers both the business and technical needs* that may not be acceptable to others. The definition also may not include all necessary concerns. Additional requirements are in order.

DEFINITION REQUIREMENTS

Before defining systems engineering, it is necessary to answer a basic question - "What are the requirements related to a definition?" The requirement areas for a definition are: 1) the function of the definition(s), 2) how well the function must be accomplished, 3) where and under what conditions and by whom the definition will be used, 4) what constraints affect acceptability of the definition. These requirements are defined in Table 1.

Table 1 - Definition Requirements

Requirement Area	*Requirement*
1. Function of definition(s)	Communicate meaning to the stakeholder set defined in Table 2
2. How well the function must be accomplished	(see Table 2)
3. Where and under what conditions the definition(s) will be used and by whom	(see Table 2)
4. Important constraints	a. Definition(s) must be acceptable to stakeholders. b. Definition(s) must be short and comprehensive. c. Definitions(s) must be scalable to large and small applications. d. Definition(s) must be flexible so as to be applicable to new and incremental developments, as well as modifications and changes to existing systems or products under development. e. The definition(s) must not ignore the already established definitions provided in published standards and maturity evaluation tools.

CRITERIA FOR AN ACCEPTABLE DEFINITION

The criteria areas for a definition of systems engineering are: 1) the function(s) of systems engineering, 2) how well the systems engineering function(s) must be accomplished, 3) where and under what conditions and by whom systems engineering is accomplished, 4) what process and design constraints affect systems engineering, 5) who the customers for a definition of systems engineering are, and 6) who uses a systems engineering definition and how. The criteria for an acceptable definition of systems engineering are provided in Table 3.

In addition to the criteria of Table 3, systems engineering is concerned with any man made system that has a purpose and is made up of operational and enabling products which are composed of one or more of the following elements - hardware, software, people (humanware), data, materials, services, techniques, and facilities [EIA 1994]. The products and processes of the system must be developed, manufactured, verified/tested, deployed/installed, operated, and supported. Additionally operators and users of the products must be trained, and spent products must be properly dispositioned along with by-products from manufacturing, test, deployment, training, and support processes.

To better understand the functions of systems engineering the main function "engineer a complex system" can be decomposed into its subfunctions. The main two subfunctions are *develop system operational and enabling products* and *manage technical development*.

Product development requires defining system requirements, analyzing the functions of the system, designing physical solutions, and verifying that selected physical solution satisfies stakeholder requirements [IEEE 1995; Grady 1995]. Management of the technical development requires planning, organizing, and controlling the technical efforts.

Table 2 - Stakeholder Requirements[57]

Stakeholder	Example Requirements
1. Enterprises and their constituents	Vision Statement - "To be recognized for our premier systems engineering" Definition(s) must convey what they mean by this statement.
2. Trainers and consultants	When an instructor or consultant displays the definition(s) to a group made up of program/ project managers, or integrated product team members, or system engineers, it stands alone without the instructor/consultant having to take additional time to explain what the definition means and motivates the audience to be attentive and willing to listen to that which follows in the presentation, be it one hour or five days. It should be able to focus interest and provide justification for the presentation.
3. Educators	When used by a college professor, it helps set the direction and/or focus of a full semester course or degree program, and help justify the subject matter.
4. INCOSE potential members	When used in an INCOSE advertising or other command media, any potential class of members (student, associate (non-engineers), or member) will want to become involved with the Council. It should not only attract system engineers as defined by the profession or discipline.
5. INCOSE decision makers	The definition(s) must be acceptable to INCOSE decision making bodies (Technical Board, Board of Directors, Corporate Advisory Board (CAB)). Definition(s) must represent the diverse philosophies previously referenced in the introduction.
6. Executives	When an executive (person not directly in charge of a systems engineering organization) is asked to have his or her enterprise join as a CAB member, the definition(s) should enable, not hinder, the willingness to pay $10,000 for membership and support of INCOSE. Definition(s) should also motivate willingness to serve on the CAB.
7. Integrated product team members	Definition(s) should motivate involvement in systems engineering activities.
8. Manager of system engineers	When a manager of system engineers (the discipline/profession) approaches his/her budget line supervisor (e.g., non-system engineer Vice President, Program Manager, Director), the definition(s) should enable, not hinder, the granting of funds.

---

[57] This list of stakeholders is meant to represent an exemplary, not complete, set.

Table 3 - Requirements for a Definition of "Systems Engineering"

Criteria	Requirement
1. Function of systems engineering	a. To "engineer a complex system," or b. To "develop product solution," or c. To "produce product design."
2. How well the systems engineering function must be accomplished	Resulting products and processes must: a. satisfy tasking activity/acquirer/user/ customer requirements, b. meet organizational commitments, and c. meet public expectations.
3. Where and under what conditions and by whom systems engineering is accomplished	a. Systems engineering may be accomplished within commercial or government enterprises. b. Systems engineering may be accomplished under competitive or non-competitive conditions, under limited resource availability, and under contractual or non-contractual conditions. c. Systems engineering tasks will be accomplished by those persons involved with the engineering of a system to include engineering and non-engineering personnel; technical and non-technical personnel; and individuals or teams of individuals.
4. Process and design constraints	a. Should not define what a person does or what a systems engineering organization does. b. Should not limit systems engineering definition to a part of systems engineering - e.g., modeling, analysis, methodology, tool set. c. Should not define what a system engineer does or is. d. Should not justify or specify the profession of a system engineer or systems engineering whether looked at as a career field or academic area. e. Should not make systems engineering a discipline, since it is practiced by multiple disciplines. f. Should require a disciplined (ordered) practice. g. Should be scalable to any size project. h. Should encompass the total system, e.g., any item of the system whether at the system, subsystem, or lower level layer.

**Table 3 - Requirements for a Definition of "Systems Engineering"
(continued)**

5. Customers for a definition of systems engineering	Included are: a. Members and potential individual and corporate members of INCOSE. b. Enterprises responsible for the engineering of systems. c. Faculty and students of systems engineering programs. d. Practicing system engineers. e. Members of multidisciplinary teams.
6. Users of a systems engineering definition and how they will use it	Refer to Table 2

DEFINITIONS OF SYSTEMS ENGINEERING

With the above considerations, appropriate definitions can be synthesized. Two types of definitions are possible - what systems engineering is and what systems engineering does. The first type definition is more generalized and formal, like one would find in a dictionary. The second is the operational one. It explains what systems engineering does and would best be used in selling INCOSE and in systems engineering training classes.

What Systems Engineering Is

A definition of "what systems engineering is":

Systems engineering is an interdisciplinary, comprehensive approach to solving complex system problems and satisfying stakeholder requirements.

By interdisciplinary it is meant that systems engineering of complex systems is not possible by a single individual. It requires persons from a variety of engineering and non-engineering specialties and functional areas to provide their skills and knowledge in an integrated manner so that the system solution incorporates all required attributes efficiently and effectively. An approach is comprehensive if it is:

1. well defined (each activity is defined in terms of inputs and outputs, entrance and exit criteria, work products, activity ordering, control mechanisms, and progress metrics);

2. well managed (appropriate planning, organizing, and control activities included);

3. scalable (tailorable to different sized systems); and

4. disciplined (includes mechanisms to ensure that activities are adequately followed).

Complex system problems include new system/product developments, incremental product developments, family of products development, modifications, and engineering changes. Solving the problem includes creating the system solution.

Requirements of stakeholders reflect needs, wants, expectations, desires, what they will be happy with, and what they are capable of. Requirements include: what the product is to accomplish, performance features, price, life-cycle cost, or delivery times of an external customer (e.g., developer or consumer); social responsibilities to the public (e.g., legal, regulatory, environmental); and organizational commitments (e.g., budget, schedule, resource use).

What Systems Engineering Does

A definition of "what systems engineering does":

Systems engineering evolves the definition of a product item through distinct life-cycle development activities — need or opportunity analysis, concept definition, system definition, preliminary design, detailed design, and evaluation. Upon completion of these primary development activities, there are post-development activities for systems engineering to perform. During the post-development period of a product, the systems engineering primary development activities of opportunity analysis, concept definition, system definition, preliminary design, detailed design, and evaluation are accomplished to:

a. *correct product design deficiencies discovered during production, production test, deployment/installation, training, operation, support, and disposition;*

b. *improve deployed products to make them more competitive, secure, safe, and/or marketable; and*

c. *make modifications to the product or its related life-cycle processes such as production, test, deployment, installation, training, support, or disposition.*

Systems engineering also encompasses the planning, organizing, control, and implementation activities of the systems engineering process defined in EIA/IS-632 and IEEE 1220-1994. The activities include:

a. *analysis of the development problem (analysis of functional and performance requirements);*

b. *synthesizing design solutions for the development problem;*

c. *assessment of alternative solutions and selection of the best set of balanced requirements, functions, and product and process solutions;*

d. *verifying that the selected physical solution meets requirements derived from the analysis of the development problem;*

e. *capturing all design results and rationale and data on supporting models, tools, assessments and trade studies, and plans for the management and implementation of systems engineering activities;*

f. *information, data, and configuration management;*

g. *requirements, interface, and change management;*

h. *risk management;*

i. *performance-based measurements such as technical performance measurement, earned value, and design reviews and audits; and*

j. *planning and organizing to accomplish the activities of a. through i.*

This lengthy description of what systems engineering does is the essence of what it takes to engineer a complex system, the stated functional requirement of a systems engineering definition (Table 3, requirement #1a). This definition of systems engineering is designed to meet the performance requirement of Table 3, requirement #2 in that it captures the total spectrum of stakeholder requirements. Satisfaction of those requirements will of course depend on how well the activities of systems engineering are accomplished through multidisciplinary teamwork.

There is nothing in the above definitions that limits application to either commercial or to government, to competitive or non-competitive conditions, to precedented or unprecedented systems, or to other condition requirements established in Table 3, requirement #3. Likewise, the definitions meet the constraints of a systems engineering definition (Table 3, requirement #4). For example, the above definitions of systems engineering are much broader than what a system engineer does, although the system engineer is involved with most of the systems engineering activities in one way or another. Also, the definition of what systems engineering does involves the systems engineering process described in current systems engineering standards, thus a disciplined practice is involved. The definitions are sufficiently broad to meet the needs of listed customers in Table 3, requirement #5 for a systems engineering definition. Finally, these definitions are capable of meeting the stakeholder requirements of Table 2 as required by Table 3, requirement #6.

The detailed analysis of comparing these definitions to a particular application using the requirements of Tables 3 is left to the reader.

CONCLUSION

Systems engineering does what systems engineering is. That is what the definitions provided above illustrate.

Does this end the argument of what is systems engineering? Of course it won't, nor should it. As a participating contributor to the current systems engineering standards, I have derived herein different definitions than provided in the standards. In the next work I accomplish I may well produce yet another definition. This should not be considered unreasonable nor that those in this paper or the standards are wrong. Definitions in standards, as well as in most papers and books, provide definitions that apply to what the author is trying to say. Authors don't necessarily mean that the definition they give is the only one, but it is meant to be one that will, if understood, help the reader understand the author's work.

If you disagree with the definition an author gives, or the definition doesn't meet the criteria established in this paper, you may conclude that the remainder of the book or paper isn't going to do much for you. Don't give up. Read further. You may actually learn new ideas from looking at the subject from a different perspective.

Some readers will continue to grasp for a universal definition. Perhaps one actually will exist some day (how boring). But I have long accepted the axiom that engineers must agree to disagree on certain issues so that they can progress on to bigger issues.

Nevertheless, any definition must be based on some set of reasonable criteria and must not violate the purpose of the organization, book, or paper. Tables 1 through 3 of this paper will help.

One last conclusion. If one is to produce a definition of systems engineering for whatever use or context, application of a systems engineering approach is required. Such an approach was applied in this paper. You know what? It works.

REFERENCES

Alessi, R. S., et al, "The Foundation of Systems Engineering." Proceedings Fifth Annual International Symposium National Council on Systems Engineering (St. Louis, MO, July 22-26,1995). pp. 843-850.

Chase, W. P., Management of System Engineering. Robert E. Krieger Publishing Co., Inc., Malabar, FL, 1974.

EIA Interim Standard Systems Engineering. EIA/IS-632, Electronic Industries Association, Arlington, VA, December 1994.

Goode, H. H. and Machol, R. E., Systems Engineering. McGraw Hill, New York, 1959.

Grady, J. O., "The Necessity of Logical Continuity." Proceedings Fifth Annual International Symposium National Council on Systems Engineering (St. Louis, MO, July 22-26, 1995). pp. 691-699.

IEEE Trial-Use Standard for Application and Management of the Systems Engineering Process. IEEE 1220-1994, Institute of Electrical and Electronics Engineers, New York, February 1995.

Woods, T. W., "First Principles: Systems and Their Analysis." Proceedings Third Annual International Symposium National Council on Systems Engineering (Arlington, VA, August 1993). pp. 41-46.

BIOGRAPHY

Jerry is a Director-At-Large of INCOSE. He was one of the founders and served as the first elected president. Currently he is a principal of Systems Management international, a consulting and training company. His former positions include: Pilot and R&D Manager in the US Air Force, Consultant/Project Manager for Joint Cruise Missile Program Office testing, Business School Dean at Oklahoma Baptist University, Director of Graduate Technical and Engineering Management Programs at the University of Maryland, and Faculty Member in Systems Engineering at the Defense Systems Management College. Jerry was a key author of both commercial and military standards on systems engineering and currently serves on the EIA-632 Technical Committee and represents INCOSE on the JTC1/SC7/WG7 preparing an ISO systems standard.

Systems Engineering Is More Than Just A Process

Richard Harwell
Lockheed Martin Aeronautical Systems (D/73-D2)
86 S. Cobb Drive
Marietta GA 30063-6085
e-mail: rharwell@mindspring.com

Abstract. The systems engineering process is a roadmap, a pathway to help us achieve our goals. As with systems engineering tools, the process assists - but is not a substitute for "getting the job done". Unlike many processes, systems engineering must have a proper environment to flourish - *in fact, it's an inherent element of the application of the process.*

Success In Applying Systems Engineering Varies

In today's increasingly competitive environment, many companies are turning to systems engineering to improve their responsiveness to stakeholder (customer, user, buyer, the media, the public) expectations. Many succeed, but some realize no apparent benefit from the use of the systems engineering process. Unlike many of the process fads we have been subjected to in recent times, the systems engineering process is very well documented in public media - and there are many companies known for their successes based on its application. So why the difficulty in transferring its successful application from one operation to another? Because companies often overlook the critical fact that systems engineering is as *much a way of thinking and operating* as it is a process.

Application of systems engineering requires a commitment to make it work, to avoid the easy solution, and to stand firm when necessary. This is in addition to the need for its implementers to possess an innate understanding of the technical elements of a project and how those elements should be integrated. With this complexity, it's no wonder that implementation difficulties abound - most of which can probably be traced to one of several concerns:

- insufficient management commitment
- insufficient up-front funding
- insufficient teamwork and team communications
- insufficient systems engineering skill levels
- hesitation to challenge (within the system) requirements and decisions, and
- inadequate systems engineering infrastructure.

So How Do We Establish "A Proper Environment?"

The fundamental element is the firm commitment of all participants, from the most senior member of management to the new hire at his or her work station. Application of the systems engineering process *requires a major cultural change* from the top down. This is true whether or not the value-added aspects of the process are known and understood. For example, it requires management to consistently demonstrate a firm belief in the systems engineering approach *–or the team will recognize it's just another slogan.*

Management's lack of commitment may be demonstrated in many ways:

- minimal direct involvement of the systems engineering leader in management strategy sessions,

- burying the systems engineering team in the organizational structure, and

- a limited budget without a commensurate reduction in responsibilities.

These difficulties are not deliberate attempts to undermine application of the systems engineering process, but rather *a lack of understanding of the power of a structured development environment* - and limited bottom-line oriented documentation that its application really works!

Of course, this is easier said than done, since it requires major shifts in funding from project downstream to project start as a means of properly funding the critical up-front systems engineering effort. (For DOD contractors, this also means that some funding will need to shift from contract to bid & proposal for the same reasons.) This shift in funding will enable the systems engineering team to identify, validate, and integrate the initial requirements set necessary to provide a foundation from which the remainder of the project teams can start their effort.

Another problem in achieving this commitment is the continued absence of effective substantiating information to warrant management's feeling comfortable with the necessary changes. It's also difficult to promote the enthusiasm necessary to change the culture throughout the project team. (On a similar note, how many of us are willing to start the day by developing a "to-do list" and then following it during the day - allowing, of course, for adjustments as conditions merit?)

And that's the major concern. The company's management team must believe the systems engineering process will work, or they will not commit to it.

Systems Engineering Is More Than Deriving Requirements!

Although a management commitment to (and belief in) the systems engineering process is a major step in implementing the associated environment, the effort doesn't stop there. The marketplace has changed. Development processes used in the past do not enable a company to meet today's complex challenges *and* today's informed stakeholder base. It's no longer sufficient to advise a "customer" that we know what is needed and then proceed with a preconceived development goal (or one that meets a company's singular view of the marketplace). Today's customer is a partner in the development process, and a structured and disciplined environment is essential to assuring the capture and implementation of requirements which enables the team to meet expectations.

The Systems Engineering Team Must Be A Partner In The Development Process As Well

All of us resist change, and the effective application of the systems engineering process is a major change in the way a company operates. Simply having a team of systems engineers analyze stakeholder requirements and derive lower level requirements is not sufficient for a company intent on adopting the "systems engineering way of doing business." But it is an often used means of meeting a customer requirement to "do systems engineering" while avoiding the difficulty of training and implementing the project team in a new culture. Responding to a contract requirement as the reason for implementing systems engineering does not embrace it as a way of doing business!

It's essential, therefore, that project managers also believe in the effectiveness of applying a disciplined process, and avoid the practice of proceeding from stakeholder need to preconceived design. The classic conflict (and separation) between a systems engineering team and the development team needs to be avoided. In fact, they should not be separate teams, but one project team operating to a single goal - meeting stakeholder expectations in developing a product. This means including the systems engineering team as part of the decision management process for technical effort.

Risk items, for example, could be addressed in a technical management council (or whatever the management decision/issues resolution body is called) rather than in a separate risk management board. Technical reviews could be jointly planned. Required budget adjustments could be based on the needs of the entire team, rather than learned responses (i.e., selecting visible targets rather than assessing change impact).

Integration of the systems engineering team is a two-way endeavor. It's emphasis should be on helping the project manager to match project objectives for the systems engineering team to the right budget - and to avoid over commitment when available skills and budget contraindicate the ability to deliver.

The Systems Engineering Leader Sets The Standard For An Effective Environment

One of the primary roles of the systems engineering leader is to sense, anticipate, and prepare the project team for change. In short, to understand the nuances of the project's objectives and technology and detect (early) the subtle shifts which impact projects. This entails "sticking your neck out!" The systems engineering leader must be willing to risk being wrong, rather than risking not alerting the project to an impending difficulty and guiding the corrective response.

The Bottom Line - project needs, not career self-promotion, govern!

In fact, it may be true that one can identify a "good" systems engineering leader by the number of scars on the leader's neck. Think of the career opportunities it provides (and the chance to learn how others implement systems engineering)!

The Systems Engineering Team Must Also Make A Commitment

The systems engineering team (whether applied to a project as a functional organization, an analysis and integration team, or integrated in multidisciplinary teams) must also make a commitment to ensuring the right environment exists. They must:

- establish the vision,
- be firm,
- be a value-added team player,
- demonstrate inherent integrity, and
- be innovative and maintain flexibility.

Establish the Vision. The systems engineering team needs to have a *focused, disciplined vision* of where the project team is going and how it's going to get there. They must have the ability to sense and acknowledge changes, determine their impact on the vision, and rapidly adjust to them for the benefit of the project team. This means an ability to discern patterns, disjoints, anomalies, and potential conflicts more rapidly than the team at large.

The systems engineering team needs to maintain a "system view" (a view of the product and its interrelated elements, and the environment in which it operates) coupled with an understanding of both technical and acquisition systems knowledge; i.e.:

- How does the product affect its environment?
- How is the product affected by its environment?
- How will the politics of the environment affect the requirements?
- Can the technical elements be performed within the acquisition imposed constraints?

Be firm. The systems engineer needs to question everything (although this runs contrary to a person's normal manner of performance). "Why" cannot be asked too often. Don't be satisfied until it "feels right". When it feels wrong, find out why. Investigate, document the concerns and the answers, and define the problem and potential solutions for the project leadership. *The systems engineering team must act as the "Devil's Advocate" operating within the system - a constant questioning of whether or not the project team has the right focus or is doing the right thing.*

Suggested areas to consider are:

- Is it technically achievable for the price and schedule?

  If not, drive a stake in the ground for one of the three variables and adjust, as necessary, the other two before work starts.

- Do we have the right team? The right resources?

  If not, don't start until we do - or change what is required.

- Do we have the right requirements?

  Although it seems contrary to the systems engineer's mission, requirements must be challenged at all times to ensure they capture the stakeholders' intent.

Be A Value-Added Team Player. As stated previously, the systems engineering team must be an integrated unit within the overall project team. It must also be a team player at all times. The systems engineering team must be able to demonstrate its value-added benefit to the project, whether its removing obstacles from the teams path, anticipating changes, eliminating unneeded documentation, or simply "rolling up its collective sleeves" and assisting the project team in a time of need. Most importantly, the systems engineering team should not permit itself to be an isolated, independent member of the project effort. As with any team, if the systems engineering team doesn't add value, why have it on the project?

Demonstrate Inherent Integrity? Systems engineers must demonstrate inherent integrity at all times. Bias destroys the ability to achieve a balanced solution (it's difficult to walk away from a favorite approach, even if it's not the best for the project). The systems engineering team cannot afford to show favorites, or their impartiality will always be questioned. Good communications is a factor in demonstrating this essential element. Provide constant and frequent feedback to all members of the team on project decisions and directions. Ensure affected members of the team always receive communications affecting their ability to perform.

Be Innovative and Maintain Flexibility. The objective of the systems engineering process is relatively simple, but (as a discipline) we rarely apply it in a simple manner. In many respects, we have adopted an entrenched, (i.e., complex paper-based) methodology as the process - failing to separate the methodology from the process. The needs of the project drive the methodology for applying the systems engineering process.

For example:

- Application of the systems engineering process on a research project will be different than that on a complex weapon system, and

- Application of the systems engineering process on a prototype vehicle will be different than that on the final product.

This failure of systems engineering teams to separate process from implementing methodology may be one of the major reasons many project managers resist its application on their programs. They are aware only of the complex, cost-driving, and sometimes restrictive methodologies passed off as the systems engineering process. For example, most consider the Systems Engineering Management Plan (SEMP) as a thick, unread document that represents a method for development not used by the project team. In other words - an excellent door stop! But of little value otherwise. Yet the SEMP (or some form of an engineering plan) should:

1) identify and provide insight into the problem facing the project team,

2) identify the systems engineering process to be applied to solving that problem,

3) identify the methodologies and tools to be used in applying the systems engineering process,

4) identify how the team will be structured and how the team will implement the identified methodologies and tools (and identify what training is needed), and

5) identify the metrics that will enable the team to know when the solution to the problem has been achieved.

Each project will present a different challenge and a different method for implementing the systems engineering process - and that requires innovative thinking on the part of the systems engineering team. The focus must be on adapting to the needs of the project so the project can solve the stakeholder's needs.

Systems Engineering Process Cannot Be Implemented In A Vacuum

Simply establishing a systems engineering team and charging that team with implementation of the process will not achieve the desired objectives. This approach will result in systems engineering being an adjunct to the project (and an expensive one at that). The successful process is not for an isolated team "doing systems engineering." It is an environment (an infrastructure) in which the company, the project team, and team members need to operate on a daily basis. *It defines how the organization discerns a problem, how it approaches the development of a solution to that problem, and how it implements the plan which solves that problem.*

To accomplish this, all participants in the process must:

1) Maintain a focused, disciplined vision of where they are going,

2) Tenaciously champion (and, when necessary, challenge) stakeholder objectives,

3) Provide continual, solid, truthful feedback to all members of the team on a consistent basis,

4) Shape the process, operate it as team members, and adhere to it,

5) Make these principles (and the process) the core of the value system.

The development system is fragile. If all do not share the commitment, failure to meet the project objectives is inevitable.

In short—It's hard to paddle upstream if the canoe is not in the water!

BIOGRAPHY

Richard Harwell is a Staff Specialist in Systems Engineering at Lockheed Martin Aeronautical Systems. He is the Co-Chair of the INCOSE Systems Engineering Management Technical Committee and is the founder and Chair of its Systems Engineering Management Methods Working Group. He is also systems engineering lead on LMAS' RMPA program, Principal Investigator of LMAS' "Systems Engineering Technologies Development IR&D Project, and developer/instructor of the LMAS Technical Institute's Systems Engineering courses. Recent assignments include establishing and managing the Systems Engineering Directorate for Lockheed Martin Idaho Technologies and performing as the Integrated Requirements/Risk Manager for the Grumman-Boeing-Lockheed AX team. Rich is an active member of the AIAA Systems Engineering Technical Committee and the EIA G-47 Committee. He currently serves on the EIA-632 Technical Committee

appendix B

The multi-disciplinary team concept

"Clearly no group can as an entity create ideas. Only individuals can do this. A group of individuals may, however, stimulate one another in the creation of ideas."
—Estill I. Green, VP Bell Labs

The Multi-Disciplinary Team Concept is an integral part of the SE process as defined in this book. Teaming refers to the use of multi-disciplined design teams working throughout the project life cycle. Teaming is an essential element of Concurrent Engineering (CE) and Integrated Product Development (IPD). The key to successful teams is for all team members to focus on the entire system and not just their functional area. Chapter 11 also discusses how to organize the product teams relative to the system architecture.

The team concept may be implemented in your organization by the actions of Integrated Product Teams (IPTs), also known as system integration teams or integrated product development teams. The specific makeup and leadership within IPTs may change depending on the phase of product development, but will normally consist of a mix of systems engineers, traditional design engineers, "specialty" engineers, test engineers, production engineers, logistics engineers, etc. In the early stages of product development, the team may be led by engineers skilled in functional analysis. As the product development process progresses, the lead will shift to traditional design engineers, production engineers, deployment and support engineers as appropriate. The exact composition of teams will vary and must be tailored to the needs of the specific project.

The purpose and rationale behind the use of teams in the SE process is to avoid the shortcomings experienced in a sequential development approach. For complex products and systems, the sequential approach has led to segmented, inefficient and sub-optimized systems. By bringing production and support engineers into early design efforts, your organization will be able to consider cost, manufacturing and support factors prior to selecting a product design and its required processes.

When multi-disciplinary teams are not used or not used effectively, product designs are often difficult or expensive to produce, cannot be effectively supported, operated or maintained, or require major modifications after production.

The SE process, by effectively employing multi-disciplined teams throughout the project development cycle, including design, risk analysis, and trade studies, provides balance between competing engineering considerations to arrive at an optimum product/process solution.

The use of teams in a CE approach to product and process development will provide your organization with the best SE process to satisfy customer requirements. The CE approach will involve greater up front participation by production, test and evaluation, support, and other engineering disciplines; however, the payback is a more competitive bid, and lower production and life cycle costs.

appendix C

DOD acquisition life cycle

"A great wind is blowing, and that gives you either imagination or a headache."

—Catherine the Great

Figure C-1 shows the phases of the Department of Defense (DOD) acquisition life cycle, the key milestone decisions that occur before each phase, and key events and reports that occur during the life cycle. A brief description of each phase follows.

Concept Exploration and Definition Phase

The systems engineering activity during the concept exploration phase is translating the users' needs into alternative design concepts through functional analysis, synthesis, and trade-off analysis. This is accomplished by exploring various alternatives to satisfy these needs, defining the most promising system concept(s), and developing supporting analyses and information to include identifying high risk areas and risk abatement approaches.

Demonstration and Validation (Dem/Val) Phase[58]

The objectives of the Demonstration/Validation Phase are to prove that technologies and processes critical to the most promising system concept(s) are understood and attainable, and to better define the

---

[58]The name of this phase has recently been changed to "Program Definition and Risk Reduction." The goals and objectives are essentially the same, but with a greater emphasis on reducing or eliminating the major technical and programmatic risks before proceeding to full scale development in the next phase.

critical design characteristics and expected capabilities of the system concept(s). System level requirements are defined and refined, major system configurations are identified and analyzed, and risk abatement is pursued for subsystems, materials and manufacturing capabilities.

The government product of this phase is a Systems Requirements Document and all documentation necessary to establish a functional baseline with many of the functional requirements allocated with their associated constraints. This document does not constitute selection of a specific design, but rather establishes system level requirements, employment and deployment environments, and constraints based on identification of feasible, affordable ranges of cost and system effectiveness. Proper identification of requirements, in performance terms, is essential to an effective acquisition strategy, since real competition requires a Systems Requirements Document that can be met by more than one design concept. The system specification is used by the government and contractor to mutually define, from a performance perspective, the system level operational performance, capability, functional and supportability requirements and the methods for their verification. The results are documented in an offeror unique system specification.

During the systems engineering synthesis, required configuration items (CIs) are identified. The process includes trade-off analyses to ensure that the system will satisfy the development specification performance requirements with the best possible balance of LCC, schedule, system effectiveness, manufacturability, and supportability.

Elements of the proposed system are continually assessed to identify areas of technical uncertainty that must be resolved in later project phases (risk assessment). Critical components, manufacturing processes, and training and support products should be rapidly prototyped to reduce risk. At the end of the Dem/Val phase (or early in the Engineering and Manufacturing Development phase) agreement is reached on the system level requirements, and the major configuration items are defined. The contractor develops a set of development specifications for the major configuration items. These subsystem/equipment development specifications are written using MIL-PRIME development specifications. The MIL-PRIME specifications contain a non-contractual handbook which provides rationale, guidance, lessons learned and instruments necessary to tailor the requirements and verification sections of the MIL-PRIME specification for a specific application. As such, the system and subsystem development specifications provide the technical management framework for the government and contractor Integrated Product Teams.

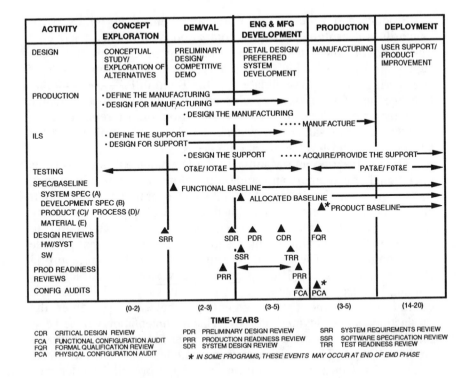

Figure C-1 *Typical Acquisition Schedule*

Engineering and Manufacturing Development (EMD) Phase

The purpose of the Engineering and Manufacturing Development phase is to complete all the activities necessary to go to full-rate production, and to field and fully support the system. This is done by completing detailed design and development of the total product (and related processes) including manufacturing, training and support, and demonstrating that all requirements are met. The Engineering and Manufacturing Development design activity is based on development specification requirements and supporting in-process event-oriented verifications using the systems engineering master schedules (SEMS). Risk is continually assessed using the technical performance measurements, SEMS success criteria, cost/schedule control system criteria, risk handling techniques and options developed and planned. Integrated Product and Process Development activities are the primary responsibility of the government and contractor integrated product teams.

Product and process definition and control activities focus primarily on resolving interface compatibility problems discovered during development testing, manufacturing process proofing and supportability verification that cut across team boundaries. Systems engineering ensures the validation of the Build-to Package, Training Package and Operation, and Support and Maintenance Package documentation. This includes verification of all requirements, including systems safety and systems activity, and completion of the subsystem/system-level verification process.

The Engineering and Manufacturing Development phase verifies operational effectiveness and suitability before deployment by testing the system in a simulation of its intended operational and support environment or in the real environment itself. Development results are reviewed to confirm that the system design meets the "exit criteria" to proceed with production, training and support activities that precede operational use. The output of Engineering and Manufacturing Development is a qualified product and verified manufacturing, training and support process that consistently yields and maintains a quality product and the documentation necessary to enter the Production and Deployment Phases. This documentation will serve as the draft product baseline until validation at Physical Configuration Audit (PCA). The documentation includes Build-to-Packages, Training Packages and Operation, and Support and Maintenance Packages.

Production/Deployment Phase

The primary objective of the Production phase is to produce and deliver an effective, fully supported system at optimal cost. In a production run where many items are to be delivered, manufacturing is usually accomplished in two segments. The first segment starts with low-rate production of initial product batches or blocks to verify processes, permit statistical process control, and establish subcontract sourcing. During the second segment, the rate increases to peak rate production as necessary changes resulting from initial operational use, experience, reviews, audits, testing, and production experience are incorporated.

Operation and Support (O/S) Phase

The O/S phase starts with deployment of the system and continues until disposal. The major activities during this period include introducing modifications and product improvements as necessary throughout deployment as well as supporting the fielded systems with items such as tools, spare parts, and technical documents.

appendix D

SE process champion's role

"Experience is a hard teacher because she gives the
test first, the lesson afterwards."
                                        —Vernon Law

The Process Champion role, when performed properly, facilitates use
of standard processes, collection of metrics to monitor the processes,
and improvement of the standard processes. Individuals selected to fill
this role will have the opportunity to make tangible, identifiable
contributions to improvement of processes.

SE Process Champions are responsible for introducing the SE
process into a project and coordinating, managing, and improving the
use of that process throughout the project's life. The roles and
responsibilities described below must be tailored to your specific
organization; the description below is generic and is only one of
several ways of implementing this role.

The SE Process Champion acts on behalf of functional
management, Project Management, and the SE PMT to carry out
process-related functions within the project. The SE Process
Champion is the point of contact that the SE PMT has with the project,
and is responsible to the Project Manager for the performance of the
process.

If more than one functional organization supports a given project,
functional managers should coordinate so that there is an appropriate
number of Process Champions for all relevant processes. In the case of
processes that span major functions (e.g., development and
manufacturing), then the process owner may be asked by the Project
Manager to provide this coordination for an individual project.

If there is only one SE Process Champion for a project, then this will typically be the Systems Engineering Manager for that project (or that person's assistant).

In coordination with the project manager, the IPT leaders and the functional management, the SE Process Champion will execute the following responsibilities:

1) Keep Project Management and the IPT leaders informed concerning the process introduction, soliciting support by communicating the benefits of the process.

2) Generate a training plan for execution of the tailored SE process, and work with project and functional management to implement it.

3) Provide liaison with the SE PMT to secure help with the process whenever required.

4) Collect the metrics required by the SE PMT, and deliver them to the appropriate stakeholders (Project Management, functional management and/or SE PMT).

5) Identify themselves to the SE PMT and, with the help of the PMT, educate themselves concerning the process, its tools, its training, and its metrics.

6) Generate the plan by which the process is introduced into the project, taking into account the specifics of the project.

7) Tailor the process, if needed, for the specific requirements of the project, minimizing the amount of tailoring as much as possible.

8) Develop the Systems Engineering Management Plan (SEMP) for the project.

9) Establish supplementary metrics, as appropriate, to measure the performance of the tailored process and allow its improvement.

10) Perhaps form a project-specific SE PMT to carry out process improvement at the project level.

11) Make specific suggestions for process improvement and deliver these to the SE PMT.

12) Coordinate and review for consistency process metrics from other functional areas on the project (e.g., software, physical/mechanical, etc.).

13) Act as the process owner of the local, tailored process, with responsibility for management and continual improvement of that process throughout the project's life. Obtain training in process improvement, if not already trained.

appendix E

Systems engineering life cycle models

"The system ... is the best that the present views and circumstances will permit."
—Alexander Hamilton

Several models exist for life cycle development of a system. Usually one or more of the following models will best fit your particular application. Quite often for large systems, a different life cycle model might be applicable for the different components due to different schedule constraints, experience levels or risk levels for each component.

<u>Waterfall Model</u>

The waterfall model is based on a top-down development approach and is an ordered set of phases that are performed in sequence. The output from one phase becomes the input to the next. The problem statement is input to the initial phase and the tested solution is output from the final phase. This model has several advantages:

a) It is an orderly, systematic model for managing the size and complexity of a system development.

b) It is a top-down development method that is compatible with top-down Functional Analysis/Allocation and bottom-up Synthesis.

c) There is some variation on how it can be used. If a portion of the system is not well understood, prototyping can be used in the requirements analysis phase. If the project is large, risk can be reduced by scheduling incremental deliveries.

The disadvantages of the waterfall model include the following:

a) It works best when the requirements can be completely
    specified before starting preliminary and detailed design.
    Prototyping during the requirements analysis phase can be used
    to reduce the risk associated with partially specified
    requirements. While this may help, requirements changes and
    clarifications are expensive to incorporate in a project using the
    waterfall model.

b) It encourages a communication gap between the customers and
    the developers. The waterfall model forces the customer to
    specify the desired system up-front and wait until the last phase
    before they can test drive it. Incremental deliveries can
    ameliorate this situation somewhat. However, once the
    requirements are specified, the customer does not see an
    incremental delivery until it has been fully developed and
    tested. At this point major changes can be very expensive.

The disadvantages of the waterfall can be significant in situations
where the customer is unable to completely specify the desired system.
At the beginning of many projects, partial solutions are necessary to
stimulate further understanding of the problem based on the effect of
partial solutions. Since downstream changes are relatively expensive
to incorporate in the waterfall model, this model tends to be at a
disadvantage when applied to a project where understanding of the
problem evolves as the solution is being developed. Fortunately, there
are alternative process models that can be applied to these types of
projects.

Evolutionary Development Model

The evolutionary development model is based on the execution of
several increments of the system products and processes. The
direction for the current increment is determined by the customer's
operational experience with the previous increment. This model is
applicable to problems where the requirements are poorly understood
and several partial solutions are needed before the requirements can be
completely specified. The disadvantage with the evolutionary
development model is the assumption that operational increments will
be flexible enough to accommodate unplanned evolutionary paths.
This flexibility is accomplished by using structured rapid prototyping
techniques and tools, and requires a high degree of customer
participation during prototyping.

This model is not appropriate in the situation where several independently evolved applications must subsequently be closely integrated. It should be noted that even though the evolutionary development model is more appropriate than the waterfall model in certain cases, the shift from one model to the other does involve certain risks. It is recommended that prototyping incremental delivery be used within the framework of the waterfall model first. This approach can be used as a migration path from the waterfall model to the evolutionary development model.

Incremental Development Model

The incremental development model is based on an organization of the development into a series of partial, prespecified increments of functional capability. This model is appropriate when early capability is needed, downstream requirements are poorly understood, or there is a need to partition the system in order to reduce risk. Also, this approach can be used in concert with the waterfall model.

Reusable Components Model

The reusable components model reverses the usual requirements-to-capabilities sequence inherent in the waterfall model. This model begins with an assessment of reusable components available, and then involves adjusting requirements whenever possible to capitalize on existing components. This model is appropriate when an organization specializes in a particular problem domain. The advantages are that reusable components tend to be very reliable and, when used correctly, they tend to decrease costs and compress schedules. The disadvantage of this model is the up-front cost required to develop an infrastructure to support reuse.

Technology Application Model

The technology application model is similar to the reusable components model since this model begins with an assessment of emerging technology, and then involves adjusting requirements whenever possible to capitalize on the emerging technology. This model is appropriate when an organization specializes in a particular problem domain. The advantages are that emerging technology can be a discriminator between our products and a competitor's and, when used correctly, it can reduce costs, reduce risks, and/or increase performance capabilities. The disadvantage of this model is the increased risk of using an unproven technology. The risk of increased schedule and cost must be balanced against the benefit of increased performance and potential cost/schedule savings.

Reverse Engineering Model

The reverse engineering model is appropriate when modifications are required to an existing system or when a failure mode analysis requires an understanding of a system's capabilities and functional behavior. It is also applicable when required to understand the behavior and capabilities of an external system which interacts with the system under development. Competitive analysis will sometimes require the application of this model to a competitor's product. The advantage of this approach is that this may be the only way to determine the capabilities of a system when that system's designers are not available, not accessible, or even hostile. Also, a competitive advantage of a competitor or an enemy may be neutralized. The disadvantage of this approach is that it is quite often difficult and expensive. However, there are several development tools becoming available which assist in reverse engineering.

Spiral Model

The spiral model is a risk-driven model and is based on repeating a basic five-stage process:

a)   Determine the objectives, alternatives and constraints;

b)   Evaluate the alternatives; identify and resolve risks;

c)   Develop and verify the next level product or process;

d)   Plan the next cycle; and

e)   Review the outcome of the cycle.

The spiral model is best viewed as a meta-model—a model that generates other models. The use of the spiral model in the early phases of a project will help determine which of the previous models is appropriate for a specific project.

appendix F

Technical reviews and audits

"Well done is better than well said."
—Benjamin Franklin

A project may conduct Programmatic, Incremental and Major Technical reviews. The appropriate reviews for a particular project must be selected during the tailoring process. Programmatic reviews usually deal with nontechnical issues such as cost and schedule status and business issues and are held periodically, perhaps monthly or quarterly. Major Technical reviews in relation to design, production and deployment activities are shown on Figure F-1.

Incremental reviews consist of three types—Subsystem, Functional and Interim System. Major Technical and Incremental reviews are described in EIA 632. The intent of the Incremental reviews is to provide sufficient structure to ensure that requirements and design decisions are made in a multi-disciplinary environment, and to provide a framework for effective planning to execute required accomplishments. The Subsystem, Functional and Interim System reviews are oriented toward ensuring that each CI has met its SEMS accomplishment criteria and contractual requirements.

Subsystem reviews are held to assure that the requirements (including interface requirements) for the subsystem have been identified, met, and balanced across all prime mission products. Each review should focus on required accomplishments for upcoming Interim System and Major Technical reviews. The Subsystem review should include assessment of risk and impacts on and by interfaces with other subsystems and systems.

Functional reviews should be conducted on the functional element of the WBS for a prime mission product with a horizontal WBS relationship. Examples of Functional reviews include: development (systems engineering), support (including disposal and deployment), training, test, and manufacturing. These reviews should assess the functional area's status in satisfying the objectives and requirements of the prime mission product and other elements at the same level of the WBS. These integrated Functional reviews should be conducted at the various levels of the WBS so that the complete system requirements and needs for functional support will be addressed for all CIs, and to ensure that risks are being managed to the level commensurate with contract phase, prior to a system-level Major Technical review.

Interim System reviews should be conducted on a CI or aggregation of CIs. Prior to a Major Technical review, all Interim System reviews should be completed to support system level objectives including resolving system level conflicts and issues from functional and subsystem teams and across other CIs, integrating results from Functional and Subsystem reviews, and assessing progress toward satisfying accomplishments required for an upcoming Major Technical review.

Major Technical reviews reflect major system development milestones. These reviews should be conducted to provide an overall assessment of the development progress of the entire system, evaluate risks and risk handling measures, and verify completion of milestone events and activities. Each Major Technical review should have well-defined entry and exit criteria developed jointly by the customer and contractor representatives. These criteria are usually documented in the SEMS.

Only those Major Technical reviews that are applicable to a particular project or project phase should be selected. The plan for the conduct of each review will be contained in the SEMP. A comprehensive set of critical accomplishments and success criteria shall be included in the SEMS. Each Major Technical review shall address:

a) Systems engineering process outputs and traceability to customer needs, requirements and objectives.

b) Product and process risks.

c) Risk management approach.

d) Cost, schedule, performance and risk trade-offs conducted.

e) Critical parameters that are design cost drivers or have a significant impact on readiness, capability and life cycle cost.

f) Major trade studies conducted to balance requirements and product and process solutions.

g) Confirmation that accomplishments in the SEMS have been completed as determined by the demonstration of satisfying all associated accomplishment criteria.

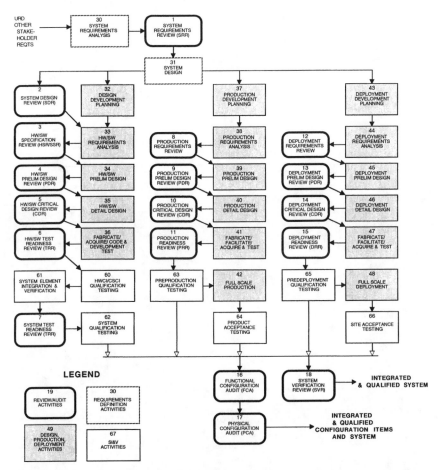

Figure F-1   *Major Technical Reviews and Audits in Relation to Design, Production and Deployment Activities*

(See Glossary for Definitions of Reviews and Audits)

Appendix G

In-process quality inspection (IPQI)

"Unrest of the spirit is a mark of life; one problem after another presents itself and in the solving of them we can find our greatest pleasure."

—Karl Menninger

In-Process Quality Inspections may be performed on documents and other outputs of the systems engineering process. These may include plans such as the SEMP or test plans, schedules, work breakdown structures, task descriptions, specifications, functional flow block diagrams, schematic block diagrams, system optical and other budgets.

The IPQI process should be followed to prepare for, carry out and report results of inspections. A form called "Inspection Defect List" may be used to record defects found during an inspection. The inspection is key to producing high quality products. In addition, inspections provide valuable information for process improvement. By reporting document defects, the System Engineering Process PMT can analyze the data to determine which defect types are most common so that process improvements can be focused in those areas. Defect codes from the following list of Systems Engineering Defect Types should be used in the appropriate column of the "Inspection Defect List." These may then be summarized and reported to the Systems Engineering PMT using a form like "Inspection Summary and Metrics."

Defect Code	Defect Type	Description
RE	Requirement	The requirement is incorrect, conflicting, or missing. Information not traceable to higher level requirements, statement of work, or contract. Error or performance budgets not properly flowed down to lower level specifications.
IF	Interface	Mismatch between interfaces of subsystems, missing interface spec.
TS	Testability	Design, plan, task, or spec does not adequately address testing. No test points. Design is hard to test. Test equipment has not been considered.
MN	Manufacturability	Design, plan, task, or spec does not adequately address manufacturability. Tooling has not been adequately addressed. Expensive or difficult to manufacture or produce.
MT	Maintainability	Design, plan, task, or spec does not adequately address maintenance. Maintenance tools and spare parts not considered. Design hard to maintain or trouble shoot. No performance monitoring or fault localization.
DC	Documentation	The documentation is incomplete, missing, or does not meet format requirements. The SEMP or other plan is missing a lower level plan or document and is needed (e.g. no Quality Plan). There are typographical errors, information is not clear and may be misinterpreted.
HF	Human Factors	Design, plan, task, or spec does not address or meet human factors or safety requirements.
OT	Other	Any other defect not described by above defects.
DE	Design	Design implementation does not provide function as needed or does not adequately meet performance requirement or design does not have enough design margin. Reliability requirements not met.
DY	Deployability	Design, plan, task, or spec does not adequately address deployment of the system. Storage and transport issues are not addressed.
FN	Function	Missing or incorrect function. Function does not meet requirements. Improper flowdown of requirements to functions.
SC	Schedule	Missing, wrong or extra date in a schedule or plan. Schedule is risky, does not have adequate margin.
TA	Task	Missing or wrong task in a Work Breakdown Structure. This could lead to a missing plan or document that is needed. Inadequately described task or task does not meet the need.

Index